THE ENGINEER IN TRANSITION TO MANAGEMENT

Learning Tool for the Engineer or Other Professional Newly Promoted to Management

OTHER IEEE PRESS BOOKS

Multidimensional Systems: Theory & Application, *Edited by N. K. Bose*
Analog Integrated Circuits, *Edited by A. B. Grebene*
Integrated-Circuit Operational Amplifiers, *Edited by R. G. Meyer*
Modern Spectrum Analysis, *Edited by D. G. Childers*
Digital Image Processing for Remote Sensing, *Edited by R. Bernstein*
Reflector Antennas, *Edited by A. W. Love*
Phase-Locked Loops & Their Application, *Edited by W. C. Lindsey and M. K. Simon*
Digital Signal Computers and Processors, *Edited by A. C. Salazar*
Systems Engineering: Methodology and Applications, *Edited by A. P. Sage*
Modern Crystal and Mechanical Filters, *Edited by D. F. Sheahan and R. A. Johnson*
Electrical Noise: Fundamentals and Sources, *Edited by M. S. Gupta*
Computer Methods in Image Analysis, *Edited by J. K. Aggarwal, R. O. Duda, and A. Rosenfeld*
Microprocessors: Fundamentals and Applications, *Edited by W. C. Lin*
Machine Recognition of Patterns, *Edited by A. K. Agrawala*
Turning Points in American Electrical History, *Edited by J. E. Brittain*
Charge-Coupled Devices: Technology and Applications, *Edited by R. Melen and D. Buss*
Spread Spectrum Techniques, *Edited by R. C. Dixon*
Electronic Switching: Central Office Systems of the World, *Edited by A. E. Joel, Jr.*
Electromagnetic Horn Antennas, *Edited by A. W. Love*
Waveform Quantization and Coding, *Edited by N. S. Jayant*
Communication Satellite Systems: An Overview of the Technology, *Edited by R. G. Gould and Y. F. Lum*
Literature Survey of Communication Satellite Systems and Technology, *Edited by J. H. W. Unger*
Solar Cells, *Edited by C. E. Backus*
Computer Networking, *Edited by R. P. Blanc and I. W. Cotton*
Communications Channels: Characterization and Behavior, *Edited by B. Goldberg*
Large-Scale Networks: Theory and Design, *Edited by F. T. Boesch*
Optical Fiber Technology, *Edited by D. Gloge*
Selected Papers in Digital Signal Processing, II, *Edited by the Digital Signal Processing Committee*
A Guide for Better Technical Presentations, *Edited by R. M. Woelfle*
Career Management: A Guide to Combating Obsolescence, *Edited by H. G. Kaufman*
Energy and Man: Technical and Social Aspects of Energy, *Edited by M. G. Morgan*
Magnetic Bubble Technology: Integrated-Circuit Magnetics for Digital Storage and Processing, *Edited by H. Chang*
Frequency Synthesis: Techniques and Applications, *Edited by J. Gorski-Popiel*
Literature in Digital Processing: Author and Permuted Title Index (Revised and Expanded Edition), *Edited by H. D. He J. F. Kaiser, and L. R. Rabiner*
Data Communications via Fading Channels, *Edited by K. Brayer*
Nonlinear Networks: Theory and Analysis, *Edited by A. N. Willson, Jr.*
Computer Communications, *Edited by P. E. Green, Jr. and R. W. Lucky*

THE ENGINEER IN TRANSITION TO MANAGEMENT

A Learning Tool for the Engineer or Other Professional Newly Promoted to Management

Irwin Gray
President
Envort-Gray Corporation

Written under the direction of the Education Committee of the IEEE Engineering Management Society: W. Maurice Kaushagen, Chairman; Newton Teixeira and Joseph P. Martino, Members.

IEEE PRESS

The Institute of Electrical and Electronics Engineers, Inc. New York

PRINTED IN THE UNITED STATES OF AMERICA

IEEE International Standard Book Numbers: Clothbound: 0-87942-111-8
Paperbound: 0-87942-112-6

Library of Congress Catalog Card Number 78-61533

Sole Worldwide Distributor (Exclusive of the IEEE):

JOHN WILEY & SONS, INC.
605 Third Ave.
New York, NY 10016

Wiley Order Numbers: Clothbound: 0-471-05212-4
Paperbound: 0-471-05213-2

CONTENTS

Foreword

TRANSITION OF THE ENGINEER—TO MANAGEMENT

The education (and early on-the-job experience) of modern engineers is an intensive and demanding one. Heavy emphasis is placed on deep technical understanding and effective application of theory and experiment in the solution of engineering problems. There is little time for broader contemplation of how the technical efforts fit into a larger context or even where the challenging problems come from.

This situation has been emphasized even more in recent years because of the increasing pace of technological change and innovation. Thus it is said that the half life of the engineer, because of expanding technology, is currently only about eight years. This encourages the younger engineer to confine career development along the established narrow path of technical know-how and updating thereof. In many instances this career development format is encouraged by the university system, by employers, and by the professional societies. During the technological emphasis of the "roaring sixties," the priorities were clearly on technical innovation. If there were problems in programs, the format was to forge ahead technically and "go back for more money" when necessary. It was as though the spiralling challenges and accomplishments would never end.

More recently we have found evidence in many directions that there is more to technological change and accomplishment than "more of the same." In fact, from the individual professional engineer's point of view, there are two prominent sets of influences:

1. Those associated with career development along managerial lines.

2. The changing technological world with the realization that technology is not necessarily good in and of itself but must carry its share of the responsibility for impact on society.

We will consider each of these two forces briefly.

As an engineer moves through his or her career, he or she has progressed in professional stature through several levels of increased technical responsibility. Inevitably these steps bring the person closer to the managerial responsibilities of the organization. During this part of a career he or she has frequently and perhaps quite naturally looked upon managerial aspects as "trouble" rather than opportunity. It is from management that one receives the message that time and money are running out. As a result, one must compromise and curtail technical ambitions. It is from management that one receives signals that his or her interests and accomplishments don't quite match the firm's needs and one must move in a different and perhaps less interesting direction. It is from management that one receives monetary rewards (sometimes beneath expectations). As a result, one develops a somewhat negative point of view towards management. "They don't appreciate what I am doing." "They don't understand that I must not compromise with ultimate technical excellence." "They don't appreciate the technical need for my large equipment budget or my extensive travel plans to professional society meetings." There is inevitably the confrontation between "they" and "me."

Thus it seems that when a managerial opportunity presents itself, the engineer is frequently not prepared to give it proper consideration either from a mind-set or from self-study and preparation for it.

It seems, however, an engineer is well qualified in many respects for managerial responsibilities even though he or she has not been involved in any formal studies of the subject. The engineer is familiar with the analytical and quantitative approach. He or she has experience in putting together complex systems and searching for optimization in their performance. He or she has learned to appreciate and use the powerful tools of the engineering world such as computers and simulations. These important and powerful capabilities need only to be extended into the managerial domain to have benefits forthcoming to the engineering professional, to management, and to society at large. Our task, then, is to bring a positive point of view to the engineer in transition to management including the contributions to be made and the rewards to be achieved.

Management is said to be both a science and an art. An engineering manager must, therefore, develop a comfortable frame of mind in the engineering/organizational/business/sociological environments that are the domain of management. Invariably people add an unpredictable and uncertain element to managerial activities. Further, there is the uncertainty of future events impinging upon management from the various interfaces with the world at large. The systems under the responsibility of management are more complex and solutions are less exact. Management presents a set of customs, vocabulary, and point of view which are somewhat unfamiliar to the trained engineer. The manager must also be comfortable with working through others rather than the ''hands on'' experiences of his or her earlier engineering career. It is a tall order for a highly trained and competent engineer to take on these additional complexities and to focus his or her capabilities and energies effectively in the new realm.

What then, should be the elements of the point of view of an engineer as he or she contemplates a transition to management? The following are meant to set the mood for thoughtful development of a new and different point of view from that developed in the earlier technical phase of his or her career:

- Management is a field with challenge and opportunity.

- Management is a field that can profit greatly from persons with an earlier engineering career.

- Management has a well-developed body of knowledge which is learnable and which needs further research, development and improvement in practice.

- Management has ''different'' and important problems to be solved.

In view of the above elements, management needs you! Management is worthy of your highest and best professional efforts. Management holds definite and rich rewards for those who can meet the challenges effectively.

—W. Maurice Kaushagen

Roadmap to the Chapters

1. INTRODUCTION

In which we discuss how the manager's job is a change in physical (visible) and mental (unfortunately, not so visible) modes of operation from that of an engineer.

S. I. Hayakwa, in his *Language in Action* tells us that, in a sense, we all live in two worlds. First, he says, "we live in the world of the happenings about us which we know at first hand." These are "the flow of events constantly passing before our senses." We call this world the *extensional world.*

The second world comes to us from books, conversation, speeches, radio, television, and newspapers. It is the *verbal world.* It is through this world we have received most of our knowledge of history and of experiences which would have been too costly or damaging for us to learn about through our senses.

The verbal world stands in relation to the "extensional world as a map does to the territory it is supposed to represent."

An engineer promoted to the position of manager stands as a child grown to adulthood. If the new adult has a verbal world "in his head which corresponds fairly closely to the extensional world that he finds around him in his widening experience, he is in relatively small danger of being shocked or hurt by what he finds, because his verbal world has told him what, more or less, to expect. He is prepared for life."

It is the purpose of this book to give the engineer "grown to supervision" a verbal map which, we hope, corresponds fairly closely to the extensional world of the manager. We hope that this book will not only give an accurate map of experience, but permit the new manager to shape the territory as well.

3. LEADERSHIP: THREE APPROACHES

4. RESPONSIBILITY FOR THE BOTTOM LINE

itiatives are producing the desired results. The wise engineering manager helps generate at least part of his own direction without waiting for definitive word from the top.

5. RED TAPE AND HOW TO DEAL WITH IT

102 Introduction

Red tape is an epithet, but from a broader viewpoint, it may be considered necessary and worthwhile. Learn how to turn it to your advantage.

102 Understanding Red Tape

102 Origins and Use

It controls information flow. Kings employed it to centralize power unto themselves.

103 Functions of Rules

The existence of rules and regulations helps the manager and shields him from some unpleasant by-products of organizational life. Learn these six ways in which rules can be beneficial.

105 Functions/Dysfunctions: Word Definitions

105 The Hierarchy of Rules and Regulations

The firm responds to external and internal pressures by generating everything from broad policy to minute directions. Here is the gamut of such guidelines and how they relate to each other.

106 External Forces

We analyze three major forces from outside the firm which generate "counter-force" rules. Comments on transitory vs. long-term considerations; do not make expedient responses which come in conflict with broader responsibilities.

Example: Dealing with customers; avoid a liability suit against the the firm—engineers, watch what you recommend.

The response to outsider queries must be filtered to avoid trouble.

Example: The "innocent" inquiry may not be so innocent.

114 Reforming the Organization

114 Overview

Look for tiny incremental improvements—not major changes if you decide to run the risks and change things. And don't be surprised if making changes is beyond your power/position/authority, or whatever, in the organization.

114 Basic Principles of Reform

Common sense is not enough. You can have the best ideas in the world for changing things. Unless you follow a careful strategic plan in how you go about working for those changes, the chances are you'll be branded disloyal and the ideas will never see the light of day (regardless of their intrinsic worth).

115 Cutting Red Tape

What is one to do when red tape truly interferes with a "critical" action? It's not as hoepless as you might thing, but neither do you "take the bull by the horns" and run. We offer suggestions on intelligent cutting.

117 Strategy of the Single Report

Elimination of paperwork involves a careful and cautious approach.

118 Your Own Red Tape

Don't ever forget that the ultimate payoff to your firm derives from the competence of your group. To check that competence is a key aspect of your engineering management job. It may require that you establish some red tape of your own—you cannot allow errors and omissions to escape and do damage to the firm. Here is, also, an example format you might want to use for a report which demonstrates the concept of completed staff work.

120 Conclusions

You may no longer consider red tape a pathology now that you have seen why and how it works. You will also be better equipped to deal with it in a constructive fashion.

6. THE MANAGER'S BASIC BOOKSHELF

122 After you finish this transition book, you should begin to read into the literature pertinent to the management field. We present a listing of books that are classics in the sense that they have been around and withstood the onslaught of "the latest findings," provide a basic foundation in concept and example, and force the reader to stretch his or her mind.

THE ENGINEER IN TRANSITION TO MANAGEMENT

A Learning Tool for the Engineer or Other Professional Newly Promoted to Management

CHAPTER 1
Introduction

Chapter 1
Introduction

WHAT THIS BOOK IS ABOUT

This is a book for the person making the transition from *performing* technical work to *managing* it. It is designed to provide operational guidance to the new manager; it fills a gap in the literature in that it tackles the problems "head on," the activities, and the hidden snares the new manager will face. This book will not "talk around" the specifics or those details which allow a philosophy or method of management to be applied on the job. It is, in this sense, not a textbook which sticks to the theory and leaves it up to the person to find a way to apply the theory. To the contrary, this book will leave it to the manager to search out the theory so that his knowledge is enhanced and his application possibilities are broadened by a larger conceptual framework.

The physical aspects of the transition are, not surprisingly, the most visible. The office, the title, the pay—these are real, measurable, and mark the differentiation between the engineer and the manager. The operational aspects also make themselves felt, but in the following ways:

(1) The time spent with people, as opposed to the time spent on the details of the technical work, increases sharply. In fact, time allocation problems become a serious limitation on what the manager can accomplish.

(2) The manager suddenly discovers that he has access to information which may not quite have been hidden before, but which now becomes positively necessary for his functioning on the job.[1] The manager, in turn, becomes a part of the management information system of the firm. Reticence among engineers is tolerated—engineers with their noses to the technical grindstone are reluctant to divert their attention from the solution of the technical problems to make reports. The manager, however, is expected to make information gathering and information dispersal up and down (and to all peers) part of his modus operandi. Indeed, a silent manager becomes suspect and the hierarchy is unforgiving of people who do not keep it informed.

(3) The nature of the rewards of the job and the satisfaction to be gained from it change. The engineer takes pride in a tough problem well mastered—the elegant solution or the neatly backed up decision which yields expected results. The output is a plan, a device, information which can be coded and used. The rewards lie in the realm of professional recognition via patents, publications, higher titles on the technical ladder, and the outward marks of accomplishment which lead to better labs, tougher problems, and more elegant solutions. The manager, by contrast, frequently suffers (mentally) the loss of a technical result more than anything else about his new job. He gains no stature from the day-to-day operation of the group; if all is smooth, no one questions him, and when the job is completed he may be the titular head of the group which did the work, but all in attendance know who did the "real" brain work on the job. The rewards must be accepted in terms of a managerial title, access to levels of the management hierarchy which the engineer rarely sees, access to information the engineer does not get, and participation in the making of decisions which affect the careers of those engineers. The manager must take satisfaction in the higher plane of personal interactions, namely, the decisions which affect more money, more resources, more accomplishments than were available to him as an engineer. His ability to make those decisions and to implement the actions to carry them out will be the source of his satisfactions and these must substitute for the technical aspects of the engineer. The new manager has gained an increment of control over the future development of his own position and a group under his authority.

(4) Finally, the nature of the reports a manager will read is vastly different from those the engineer will read. This not to say that the manager neglects the technical material the engineers has as his focus, but rather the manager will find himself more concerned with the financial aspects, the scheduling aspects, the planning aspects, or the control aspects than with the technical details. Frequently, the manager will have to read the technical summary of the reports his engineers render

[1]*Conversely, information readily accessible to him before, now becomes "not ready to show to management."*

and completely forego the immersion in the technical minutia which, formerly, would have been the heart of his review.

The mental aspects of the transition are hidden and make their impact in a more subtle way. The manager whose "mind set" is still in the technical realm will be fascinated by the technical details and give shorter shrift to the financial and people details; this manager will have problems that will get progressively worse without recognizing them until a disaster takes place. The manager whose "mind set" changes from the engineer to that of a person responsible for resources and managerial decisions will become less concerned with retaining familiarity with every technical nuance and will accept the inevitable retreat from the cutting edge of the technical realm. This manager learns to use his technical knowledge as an *auditing tool*—a base from which to question, probe, and evaluate the technical work done by others. The managerial "mind set" is the one that is opened to new perspectives and makes the mental transition equal, at least, to the physical transition.

NEEDED: A GUIDE TO THE TRANSITION PROCESS

The new manager seeking guidance on how to operate within the new job will find that most firms expect him to learn by experience. The problem is that learning by experience leads to costly mistakes, it leads to projecting the images of competence (or lack of competence) before the manager is fully aware of what he is doing, and it becomes a time of trial that is not always successful for the firm or the man himself. The cliche, "we took a good engineer and made a lousy manager," is prevalent because of the firm's expectations that people will learn by experience. This book is designed to let you learn from the experience of others. It will provide information which will let you chart the shoals on which a career might be wrecked, stalemated, or simply stunted. The material will present structures that the new manager can use to develop goals and objectives that will lead to (at least a high probability of) career success. In presenting the structures, the material will focus on feasible/operational steps—this is not a theoretical exposition of background material, but rather things to do—as specifically mentioned in the first paragraph of this chapter. Our objective is to present the new manager with means to develop personal skills for genuine growth in the managerial position.

We will accomplish our objective of helping the new manager by exposing him to a blend of theory, applied research, and the experiences of managers in the engineering profession. We will draw from the work of specialists in psychology, sociology, engineering, and the all-encompassing field of management. From these we will present a series of statements which will focus on various important aspects of the transition process. We hope to fill a gap in the managerial literature which, too often, assumes the manager is "in place," solidly in control of the job, and simply wishes to broaden his general background and concepts. By taking guidance from a solid board of engineering managers who know what difficulties are inherent in an engineer assuming a managerial position, the authors of this material appeal to the person who needs guidance at the moment that guidance is most needed—at the time of transition.

THE TRANSITION DEFINED

The transition process may be loosely defined as an exchange of mind sets, an exchange of physical parameters, and an exchange of operational parameters from those operative as an engineer to those operative as a manager. Those operative as a manager include the broader financial, personal, and information/decision aspects mentioned previously. The new manager begins using measures of success and worth to the firm which are completely different from anything done before as an engineer and which have multiple concerns instead of the technical/budget/time concerns which previously were sole operatives. The manager must, therefore, maintain his concern with the *what* and *why* instead of the *how* to become successful and respected.

SOME BACKGROUND CONCEPTS

Promotion in the firm, for an engineer, has two paths. Along one, success is measured by progress up the hierarchy and the individual assumes greater authority over a greater number of employees or larger amounts of resources. Along the second, the individual assumes more autonomy in the nature of the engineering assignments and becomes more responsible for the degree to which the results are innovative, meet the needs of the firm and its customers, and directly contribute to the technical luster of the firm's image. Advancement along the second path is called "professionalization," technical specialization, or "career specialization."

In most organizations, the career path for engineers is along the professionalization line. However, "success" is measured by progress upward in the hierarchy as an observed reward for competence. Most engineers, according to research in the field, enter industry with hierarchical goals. They strongly identify with the organization and its business goals; they want to participate in the decisions which affect their area and their careers. Further, hierarchical progress fits in with the underlying mental bent of most graduating engineers who view engineering as part of a career in a firm as opposed to a career in a specialty. A promotion into managerial ranks provides the vehicle out of less interesting (or confining) specialities into what are perceived as more glamorous fields—witness the increase in business master's degree programs versus the stability (and even decline) of professional engineering attendance at the technical seminars of conventions.

The disadvantages of an engineer moving into management often contribute to failures of many new managers. First, the promotion removes the specialist from intimate contact with the technical details resulting in the problem of instant obsolescence; this, in turn, generates fear and guilt—the engineer no longer has fall-back strength in his specialty. Second, a position in management requires skills different from those which were learned as an engineer (as becomes obvious to a new manager after only a few minutes on the job). Third, and giving rise to most of the problems, management requires dominance of personality traits and characteristics which are alien to most engineers—dealing with the diffuse, the intangible, the intractable, and with insufficient information.

Such a survey of the advantages and disadvantages of progress up the hierarchy led researchers to conclude that those who move along the lines of career or technical specialization represent individuals with a genuine interest in the technical material. They are a dedicated "core" of intensely dedicated technically oriented people who are happiest in firms with a structure that gives meaning to technical competence and technical growth. There are several firms, well known for their technical laboratories, for example, which carefully maintain mechanisms for retaining the commitment of people who have not been able to move into management and which encourage recognition of technical capabilities through great autonomy and monetary rewards.

For most engineers, however, it is only hierarchical advancement which brings increasing power, responsibility, authority, and autonomy. Such growth fits Chris Argyris's definition of increasing maturity and self actualization (self realization and growth into the person one would like to be)—greater maturity is represented by greater responsibility, influence, authority, and control over one's affairs and resources. When the engineer enters the workplace, he seeks this maturity but must demonstrate his competence in narrow specialities, under tight supervision and with little control over his resources before he will even be considered for promotion along either the technical or hierarchical ladders. If he falls into the expected line of accomplishment along the technical side of the operation, he may be labeled a career specialist and never allowed to exercise his mental set toward management; if he fails on the assignments, he kills all chances of promotion. But for those who do well on the technical side and who are then promoted into managerial positions, careful exercise of initiative and adopting the appropriate mind set should result in comfortable rewards—professionally and monetarily.

It appears to be a widely circulated myth among technical people that when an engineer flops on technical assignments he is shoved over to administrative work (writing, costing, etc.) and then, if lucky, rises up that ladder into management—all the while remaining in what is called "the technical chain." The new manager will very soon learn to reassess these old "value" judgments.

THE ORGANIZATION OF THIS BOOK

This book is designed to be a service for the engineer in transition to management. It is written under the auspices of the Engineering Management Society of the IEEE and under the direct editorial supervision of a committee of experienced engineering managers. They have seen to it that the topics, the treatment, and the writing are directed at the practical engineering manager and the reading of the material suggests operable techniques, skill development, and growth.

Each of the chapters is complete within a specific area. In some cases this may result in some duplication from section to section—this is considered more practical than to have the reader constantly turning back to previous pages to enlighten himself as to certain points. Further, in certain places distinction is made between "he" and "she" in the writing, but, in the main, the editorial "he" without regard to sexist connotations will be used.

The material is written and laid out in a new format. This format comes from the information mapping principles laid out by Robert E. Horn and is, roughly, a cross between ordinary prose and programmed learning. This format has been adopted by branches of the military and to an increasing extent by firms for technical and instructional manuals. Our reasons for use of information mapping are straightforward. Programmed learning is difficult to read. It is choppy in its presentation and makes later reference almost an impossibility. Prose, the antithesis of programming, encompasses concepts such as "transition sentences," "flow of material," and continuity which in themselves do not contain the "content." Thus, the reading for the essential points is slowed down. We sought something better. The information mapping concept comes from the idea that the "shape," or layout, or topography of the information helps transmit the information itself. Much as a map is directed to helping its readers to a specific destination, so does the mapping technique lead the reader to the specific information. The committee felt that the technique helps put across the information in an intelligent way, is quickly readable, and is useful for later reference. We hope that the readers find the layout as useful as it has been in the writing. Further, we would hope that readers will see fit to help expand the scope and depth of this work on a continuing basis. The chapters in this book are only the nucleus—we hope our readers will help us by contributing ideas, comments, and even outlines of manuscripts they would like to contribute.

CHAPTER 2

Responsibility for the Performance of Others

Chapter 2
Responsibility for the Performance of Others

INTRODUCTION TO CHAPTER

The moment that people are assigned to you, the luxury of focusing only on the circuits and components of your project is over. Now you are responsible for the proper direction of the efforts of others. Your role is to facilitate their turning out the product expected by the firm, maintaining their capabilities to turn out bigger and better products, and recognize those to be rewarded—those to be punished.

In this chapter, we try to provide some guidance to the new manager vis-a-vis two crucial areas: The management of limited time and detecting the need for action. In the chapter immediately following, we will discuss leadership. Throughout, the emphasis is on production—accomplishment of results. A person can be the brightest supervisor in the firm, have the highest morale, the best facilities, etc. If the group is not producing a steady stream of results such as drawings, designs, prototypes, solutions, and decisions which further the firm's ability to survive and grow, then the supervisor is not successful.

Success does not come by accident or seat-of-the-pants logic. It comes from knowledge of managerial systems and the use of those appropriate to the manager's style/environment. While the use of a system cannot guarantee success, it heightens the probability that results will be well above the mediocre level.

THE MANAGEMENT OF LIMITED TIME—YOUR CHOICE OF ACTIVITY

INTRODUCTION If a spectrum were drawn of work assignments by degree of activity constraint, an assembly line worker would fall at the left end and a corporation president at the right. Just where an engineer would fall is of interest here only in a relative sense, namely, a move to supervision from a technical position involves a quantum jump to the right and the seeds of trouble for too many individuals.

The jump along the spectrum is discontinuous in that: (a) a narrow focus on the work performance in the technical arena changes to dealing with focal points in the technical, financial, human, organizational, etc., areas; (b) the responsibility and authority to deal with the areas may go from nonexistant to high importance in the eyes of upper management.

Trouble sets in when the new manager loses control over daily activities in the face of new found freedoms to structure them. The new manager may make the wrong choices, become involved to an inordinate degree, or lose sight of the appropriate relationships and priorities among the focal points. Daily activities fall into patterns which may keep the manager busy, indeed may even be exhilarating and/or exhausting, but because they do not yield appropriate long-term accomplishments may spell serious trouble. The manager, therefore, needs self-guidance and a system for appropriate structuring of the day's activities to yield appropriate outcomes. This section is aimed at the manager as an individual seeking to structure the day's activities for appropriate outcome determination.

CONCEPT A manager's task choice involves elements of (a) performing it personally; (b) delegating it down; (c) shifting it to a peer; (d) pushing it upwards; (e) combinations of the preceding actions. Determination of which from (a) to (e) is appropriate for specific situations is a major element in management success.

FACT Time is the manager's scarcest resource. Expenditure for inappropriate activities leaves none for items of lasting import.

	Internal stimulus (Action)	External stimulus (Reaction)
Self planned	I	II
Self unplanned	III	IV

Figure 2.1

CONCEPT Actions may be classified in four ways as shown in Fig. 2.1. Elements of each category of action are in every job on an activity spectrum from assembly line worker to corporation president. The percentage of time spent in each category follows the spectrum from nearly all in category IV for the assembly line through the other two categories to I for the president.

The more a manager can put daily activities into category I, the more upper level management will view that manager as being on "top of his/her job."

REALITIES *(1)* Research on the expenditure of time by managers reveals that you as a manager will be spending much of your daily time in relationships (meetings, calls, discussions, interviews, stand up sessions, etc.) with dozens of people compromising, negotiating, influencing, and giving or receiving favors.

(2) Concepts of authority, power, responsibility, having meaning in the total context of a job lose relevance in situations where the job must get done in the most expeditious manner (fastest, satisfactory results, least risks, lowest costs of time and to psyche).

(3) Technical sophistication of a manager becomes of secondary importance; what counts is the ability to audit work: asking the right questions to see that the "right" work is done in the "right" way.

(4) Time spent with subordinates must be rationed carefully as to total percentage of the week it is "allowed" to consume, how it is distributed, and its relationship to time spent with peers, superiors, clients, and suppliers.

(5) "Clients" and "Suppliers" become extremely important to the success of your group. "Clients" are those to whom your group's work output goes. "Suppliers" are those from whom the group receives work (including other groups internal to the organization).

(6) The successful manager adopts a method of operating that is result oriented category I in the face of a bombardment of influences towards activity oriented category IV actions.

(7) The manager must develop an internal personal system for dealing with stresses and tensions set up by this dichotomy.

METHOD I TIME-BASE ANALYSIS CONCEPTS

Whether a time or energy consuming action will be handled personally or shifted (up, down, sideways) depends on its position on the mental grid shown in Fig. 2.2. Importance is considered in terms of its effect on:

(1) self, (2) group, (3) company.

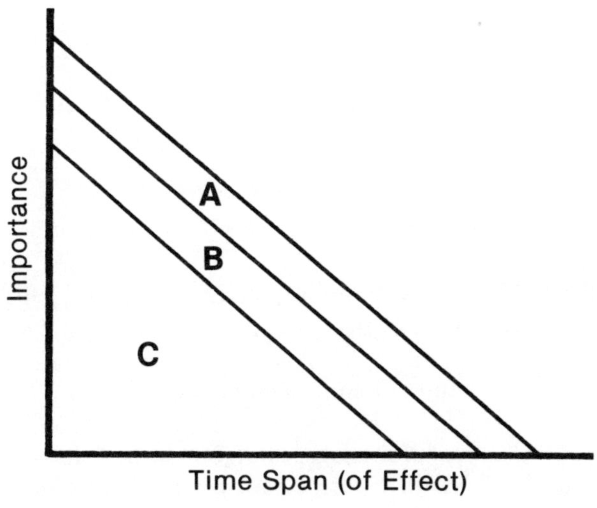

Figure 2.2

An item is "important" if it has an effect on any measure used to judge the manager. It is "very important" if it causes a significant change in this measure. It is "unimportant" if it causes little to no change in any such measure. Also, an item is "important" if it causes a change in any of the three "power" dimensions: "Performance," "Position," or "Image."

- Performance power—the ability to handle the work and resources to do it.
- Position power—the ability to command resources—supply and commitment and relationships with others holding such power.
- Image power—the view of and professional esteem in which your work is held.

Time span is the manager's best estimate over how long the matter will affect performance and image in terms of:

(a) self, (b) group, (c) company.

The chart shows three "impact" zones A, B, and C. An item will fall into a particular zone because of its long or short term effects and its importance to a measure of position, performance, and image powers. Although these are subjective judgments the manager who carefully considers an item's importance—actually asks himself/herself "how important is this?"—or judges its effective time span will make better estimates than the manager who does not internally dialogue the questions. The latter impulsively chooses or succumbs to activity patterns based on likes/dislikes or preferences/prejudices.

The manager may then choose to work only on zone A items, closely supervise others on B items, and delegate, to the greatest extent, zone C items.

Such ABC analysis is another form of Pareto[2] analysis from which we may expect that, over a reasonable period, perhaps months, 10% of all matters needing action fall into class A, 20% are B, and the rest fall into C.

EXAMPLES FOR METHOD I

(1) Accountants auditing a firm's books under tight time constraints will usually first perform an A, B, C analysis of all items/categories/books to be examined. All A items are closely examined, most B items somewhat less so, and C items are sampled.

(2) A manager confronted the following schedule items: (a) chief executive's visit; (b) important report; (c) technical hangups. There was time to take

[2]*Vilfredo Pareto (1848–1923), an Italian economist, in 1897 discovered that a large proportion of national income was attributed to only a small percentage of the total population. The principle was first applied after World War II in the area of inventory control, in which it was found that as little as 15 to 25% of the total inventory accounted for as much as 90% of the total dollar value of that inventory.*

care of only one of the three—the rest to be shifted to peers and subordinates for resolution. By training, the manager would have preferred to charge into (c), give (b) to a trusted assistant, and (a) to one of the highly personable young engineers in the lab. A little prethought indicated that (a) might have considerable impact on the image power of the manager—pushing it into the zone A category. Personal attention to this matter became paramount.

(3) An overworked manager applied ABC analysis to his desk. He then delegated a large number of C items for decision/disposition to the secretary (who was subsequently upgraded to administrative assistant). This manager suddenly discovered "enormous" blocks of time which previously had simply been "consumed."

TO USE METHOD I

In laying out your activity plan, categorize activities as to probable zone location. So-called "20 minute items" unless they represent contacts from the upper levels should be off your desk immediately for subordinates' performance. Be careful that you are sparing of what you call "A" items simply because "it has to get done." Set and stick to priorities in the "A" classes and follow up closely subordinate performance on the "B" items. The ability to distinguish between the urgent and the important is the mark of a successful manager.

METHOD II INFORMATION/ COMMITMENT ANALYSIS

The manager establishes a mental grid as shown in Fig. 2.3.

Commitment—degree to which internal motivation and action to achieve a positive outcome can be generated among those impacted by a given matter. Generally, the greater and earlier the involvement by persons concerned in the matter, the greater the degree of commitment. A higher commitment results from greater sharing of information peer-subordinate-superior and earlier involvement in a sequence of operations.

Information Level—the quantity, veracity, and validity of the information on which an action will be based. The greater the level of information the more confidence that resultant actions are correct and the less likely that pertinent facts or assumptions have been omitted.

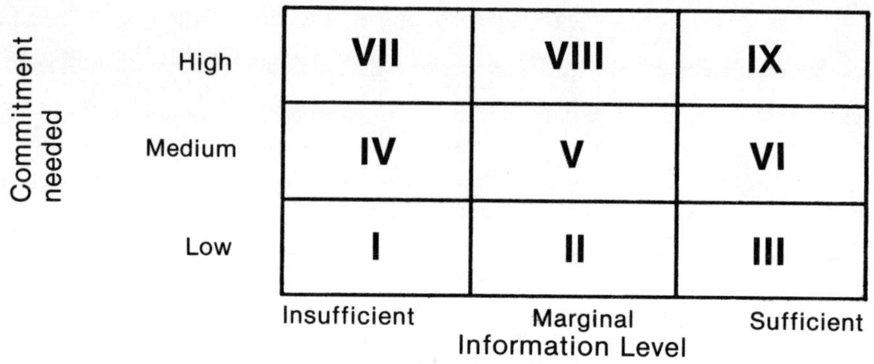

Figure 2.3

The manager dealing with categories III, VI, and IX will have to carefully consider whether taking an action on a solo basis will yield the desired commitment if this is important to the desired outcome. The temptation is for solo action which yields low commitment.

Situations in categories II, V, VIII will involve others where they will help reduce uncertainties re the quantity (have all factors of import been accounted for), validity (are we measuring what we intended to measure) or veracity (are we getting the truth behind the situation) of the information and the time is available to generate the additional information. Commitment grows with involvement. The more the manager shares, generally, the greater the return in commitment and level. Categories I, IV, VII almost always require the involvement of others because the decisionmaker has insufficient information to decide. Information must be gathered with tact and a willingness to reveal why it is needed. Where information is gathered without due regard for the interests or feelings of the donors (such as a manager's unexplained reluctance to reveal his motives) people feel used or exploited. If time pressure prevents an immediate explanation, the manager should not fail to return to the donor later.

The manager, who carefully considers the degree of commitment needed in subordinates, peers, and superiors along with the level of information at hand or to be shared, will usually be less reluctant to delegate or trust others. The mental discipline of self-dialogue and category analysis will help indicate which actions are reserved for personal attention and which should be shifted to others.

It may be noted that:

(1) Higher commitment, higher category operation, requires more time and intensity of involvement than lower category operation.
(2) Higher category operations are more revealing of weaknesses and gaps in the manager's information base—which may result in the manager's fears for his self-image.

EXAMPLES FOR METHOD II

(1) A manager with sufficient information to make a major decision, nevertheless rechecked his data with his superior. A waste of time? No! The true purpose was to predispose the superior to the decision since there was some involvement on his part in "generating" it; the manager raised his superior's degree of commitment.

(2) In laying out a schedule of activities a manager saw three areas in which information on which to act was insufficient. The tasks were delegated to people who would be impacted by the results for research, analysis, and recommendations. The manager was careful to emphasize the reservation that final action on one matter might be taken contrary to recommendations if "executive judgment" so dictated. (He correctly assumed that the "executives" might be aware of higher level, longer-term matters that could override local urgencies.)

TO USE METHOD II

Think carefully of the information you have for decisionmaking or problem solving in a particular situation. Then weigh the commitment you need from those impacted. Stop and do this, as a sort of preplan, before you take action on the matter itself. Then delegate, inquire, and involve, accordingly. Spread use of this system over more and more of your activities. Freed blocks of time should go to matters you must handle personally.

METHOD III DEFINITION LEVEL CONCEPTS

An amorphous situation or opportunity must usually be "defined" to delimit and set it apart from its context; one wishes to control only those variables pertinent to the situation and to be able to predict the most probable consequences of changes in those variables. In this context, the "level" of definition is a key concept.

Too high a level results in a definition so broad that action by multiple individuals or departments in the organization will be required to effect a result. Too low a level results in a definition in which only one action is allowed or even specified within that definition. Politicians will frequently define every problem in its broadest terms/ramifications. Thus, an action by anyone in any area has, by implication, been a result of his, the politician's, efforts to solve the problem. Conversely, anyone with a parochial interest to protect will define a problem such that his group alone can solve it and all blame falls outside his sphere of control.

The manager should aim at defining problems so that the resultant delimitation is for variables directly under the control of his/her department or immediate sphere of influence. The level of definition is raised by *why* considerations; the level is lowered by *how* considerations.

EXAMPLES FOR METHOD III

The manager is faced with falling group engineering output caused by a general feeling of "do enough to get by," "don't rock the boat," and a lack of creative spark to overcome obstacles. (Fear overcomes ambition.)

Too high a level: "What can the organization do to upgrade engineering output?" Encompasses so many potential actions and considerations, the definition of the problem by this question is meaningless.

Too low a level: "How do I push creativity into my group's head?" Implies that this is sole item to consider and can be implanted as the solution to low output.

Level to consider: Why has output fallen, and can I attack the situation by utilization of training, management by objectives (MBO), and stress methodologies, among others, open to me? Also, what do the group members say they need?

TO USE METHOD III

The item should be delegated if a problem definition incorporates the potential for actions of a subordinate within his/her sphere of responsibility.

If the situation will require action above the level of individual subordinates, those involved should meet with the supervisor to explore an overall definition. Then they should write subdefinitions at levels implying action which they can effect individually.

If the situation resists definition at a level the manager can control, it is time to seek the help of a superior, but the manager should beware of the pernicious practice of "upward delegation."

Faced with the consideration of numerous items in a busy day, the manager should examine the item/situation/decision/problem. Can the definition/delimiting of the situation be accomplished so that the range of action implied falls within a subordinate's capabilities? If the answer has a high probability of being yes, then delegate the task. Conversely, if the outlines of a situation imply broad-based action beyond the control of the manager, the matter should be examined by an upper level individual; the manager may see his/her efforts as providing supporting or clarifying information to the upper level and a willingness to assume duties arising out of the superior's definition.

The examination of a situation from its "level of definition" may result not only in more efficient utilization of the manager's time, but in better insights into the situation itself. (This is also an excellent method for communication between management levels—a superior learning/teaching device.)

METHOD IV RISK—CONSEQUENCE AND SEQUELA ANALYSIS (C & S)

This method of analyzing one's own duties is sometimes considered a derivative of Method I, Time-Base Analysis. In this method, one examines a situation from its upside (reward) and downside (cost) probabilities: The degree of exposure to losses or negative consequences represents risk and this may be acceptable or unacceptable to the manager depending on consequences and sequela.

The consequences are, simply, what will happen if an action is taken and sequela refer to the follow-on risks, responsibilities and duties which will be required if the first action is taken.

A manager examining a number of duties for delegation to subordinates should carefully examine the risks involved. Bluntly, if the subordinate "goofs up," what will happen and can this happening be appropriately dealt with? Particularly can it be dealt with without running into unpleasant consequences and the need for involving outsiders in follow-on (sequela) actions?

If risks are acceptable and C&S can be handled by subordinates, then a situation is amenable to delegation. The higher the risks the closer the manager's follow-on supervision of the matter.

The reward and probabilities all too often influence a manager to an undue degree. "I can do it better, faster, and with greater payoff" is an attitude which results in an overworked manager and underutilized subordinates. Underutilization tends to breed frustration and reduces the motivation to be a "top notch" engineer for the firm.

An overcommitted manager suffers opportunity costs each time he/she personally tackles a problem which could be done by a subordinate. Even if the subordinate's results are less than those the manager could have achieved by a personal involvement, the manager's return on time investment is highest attending to matters others cannot handle. Unfortunately, opportunity costs are invisible—they are not defined. It takes high quality time to explore them.

Managers who are overcommitted or overworked tend to exaggerate the risks and the negative consequences of a subordinate's inappropriate handling of a work situation. Further, these managers fail to appreciate their overall opportunity costs when they maintain a personal involvement in matters which could be disposed of by others.

EXAMPLE FOR METHOD IV

A manager dealing with customers can make a 5% higher return (in sales, public relations, etc.) than could a subordinate on the time involved. However, such activity results in less planning and problem resolution at the management level with resultant subordinates' idle time, underutilization, and unrectified/repeated work snafus. Cost to firm a possible 10% loss in efficiency of operations.

This manager should delegate the sales activities or customer relations to subordinates to achieve a greater *departmental* efficiency of operations.

TO USE METHOD IV

Before tackling a situation, run a mental scenario re who in your group might handle it. Assuming worst case analysis, the subordinate "bombs," can you accept consequences? Will you be able, or even care to, effect follow-on actions to rectify the situation? Can you identify a turn-around milestone where you can still correct major deviations from plan? Also, to what extent are you at risk? A good or even average employee should have a reasonable level of success. Be prepared to accept a lesser payoff than if you did it yourself; criticize or institute training where the payoff is really low. Reward where payoff is high; recognize when it is acceptable.

METHOD V AN MBO PLAN

A Personal MBO Plan. Regardless of the degree to which your organization or group operates through Management by Objectives, you may find it rewarding to operate with it personally.

BENEFITS OF MBO

The key benefits of MBO, on a personal basis, are that it:

(1) Forces you to exercise forethought on where you are going and what you will get out of it before entering the daily workaday world.
(2) Gives you benchmarks against which to judge your performance and to evaluate the actions which led to success (so you can emphasize these) and vice versa.
(3) Fosters your initiatives in the control and duration of your group because you have a clear cut sense of where you are headed. This is management by action instead of reaction.
(4) Allows introduction of management by exception principles. With clear cut objectives in mind, you as the manager will be content with limited checking on selected items as opposed to close-in constant involvement.
(5) Your actions, decisions, and initiatives will be coordinated as they are, literally, driven in a single minded manner instead of through vacillation or varying perceptions of the surrounding organization. Be careful, however, of a negative side effect: See that your use of MBO does not result in: (a) prejudging; (b) blinders; and (c) preventing your adapting to changing circumstances.
(6) Practice in using the system results in a cumulative growth of competency in dealing with all aspects of MBO.

TYPES OF OBJECTIVES

There are two types:

(1) Process or activity objectives
(2) Outcome or result objectives.

A process objective focuses on actions and decisions bearing on the internal operations of the group, department, firm operate in a more efficient manner, with better management, etc.

An outcome objective focuses on the product, the work item, the drawing, etc., which is, eventually, the productive output from which the firm generates revenue.

It is necessary to draw the distinction because too often a smoothly functioning group is equated with a successful group; a group with poor morale is equated with a poorly functioning group. While such considerations are definitely important, they are, however, parts of the process, and objectives drawn in those areas have only indirect or long term effects on the product. What is important, in keeping a firm in business, is the quality and cost of the product and the degree to which the competition is outclassed; objectives drawn relative to the product are outcome objectives and are the key considerations in judging performance success.

The formats for the two types of objectives differ only slightly; it is their differences in concept which are of key importance.

The process objective describes what you want to do to make the management operation of the firm better, to make internal processes function more smoothly, etc. The format might be:

> To [institute, change, maintain, process—an active verb] a [operation, act, method, procedure, structure] to a level of [quantity and quality measures] by [target date or over a budget cycle] at a cost of [cost of the scarce resource(s)—usually time in hours].

The outcome objective focuses on the product, the result of the act, or the information derived which furthers the economic status of the firm. Format:

> To [list accomplishment or result] at a [quality level] by [target date] at a cost of [cost of scarce resource(s)—usually time in hours].

EXAMPLES FOR METHOD V

Process Objective

To run briefing sessions, for my superior, about my group's performance at least two times a week over the next calendar year at a cost of two hours/week of my time.

Outcome Objective

To complete a report to the president about potential new business from group's engineering developments so it is comprehensive in terms of venture analysis, it is in proper format for executive decisionmaking, and has appearance calculated to forward my managerial image. Complete by June x, 19xx at a cost of 100 hours of my time.

OVERVIEW OF MBO CONCEPTS

Running your affairs without clearly delineated objectives is like firing a rifle at a hazy target area; the hits in the center will be almost by chance. Performance power is measured by hits in the center—just reaching the target area is mediocre at best.

Self-examination of one's activities (for operation by an MBO system) involves an inner dialogue on several questions:

(1) "Can I articulate what I wish to accomplish and note it on paper?" Putting your objectives on paper is a tough, demanding and extremely frustrating assignment. But it must be done if you expect to have a clearly delineated center target area.
(2) "Have I separated my process objectives from my outcome objectives, and are the process objectives clearly tied to appropriate outcomes?" Activity [or process] is impressive to your visitor. Accomplishment [outcome] is what builds performance power. Establishing a better way to keep track of your group's activities is fine and might be impressive to your boss today. Tomorrow he will be seeking the outcome of your new system and basing your performance record on that.
(3) "Have I established appropriate priorities, quality measures, and cost estimates?" If you cannot estimate, guess as realistically as possible. Keep track of your time and revise accordingly. As you keep guessing and revising according to realities, your ability to make cost estimates will improve.
(4) "Do I keep these objectives clearly in mind in decisionmaking and planning my initiatives?" Plan to accomplish your process objectives and your outcomes by orienting your daily activities toward them. Delegate as much as possible the rest of the work which might pull you off your preplanned course.
(5) "Do I revise and renew my objectives periodically in light of changing conditions?"

USING AN MBO SYSTEM

Top executives frequently complain that underlings, particularly engineering managers, have an exaggerated view of how "good" they are. This results, at least in part, because the manager may base his/her judgment on group internal operations (e.g., work hours, turnover, group activities, budgets) which are process as opposed to patents attained, technical improvements, field problems solved, etc., which are outcomes—and which yield the revenue to pay for everything.

Success consists of building a record of growing Performance Power, Position Power, Image Power. Set objectives in each of these areas and clearly show yourself the costs in time and other resources which will be required. Structure your daily activities so that you primarily handle the items pertinent to your outcome objectives and delegate or downgrade to lower priority all else.

FURTHER MBO REFERENCES

The subject of MBO fills dozens of volumes and you should refer to them for depth in the system. However, implementing a system by writing, using, revising, thinking, and evaluating only the barest concepts given here will build more personal competence than searching for perfection through study and deferred action.

George L. Morrisey, *Management by Objectives and Results.* MOR Associates, P.O. Box 5879, Buena Park, California 90622. Six cassettes and workbook. Easily the best in the field.

Westinghouse Learning Corporation, *Management by Exception.* Westinghouse Learning Corporation Training Systems Division, Customer Service Department, Westinghouse Building, Gateway Center, Pittsburgh, Pennsylvania 15222. Three cassettes and workbook.

Both reference programs are designed for the serious user—one who intends to put the ideas to use and is not just a dilettante browsing through books for something useful; do the latter in libraries. Their costs, about sixty dollars each, are not to be lightly regarded, but neither is their content. Write to publishers for exact pricing.

EXAMPLE OF NON-MBO

Jones sits in office waiting for salesmen, subordinates, visitors, superiors, and phones to determine activity pattern. Every item gets rush, immediate response treatment; where a matter cannot be handled "on the spot," Jones's subordinates immediately are detailed accordingly. Jones goes home at night dead tired and unhappy over the lack of any real accomplishment resulting from all that activity. "All I did was put out a lot of fires." "Is that really my job?"

Comment: That is the job of a fireman. Firemen (or firefighters) get paid for putting out fires. Jones is being paid to do engineering management, not to be a fireman.

GENERALIZATION

No one method in itself is a panacea or the formula by which success is achieved. Indeed, the intelligent manager will pick and choose relevant parts for his or her own use and fashion a coherent personal style accordingly.

The key items to remember, however, are that managing your own daily activity should not be a haphazard process. To the maximum degree possible, seize control of your scarcest resource, time, and orient your activities towards building a power structure (see Fig. 2.4).

Periodically, evaluate your results by yourself, with spouse or trusted confidant, and with superior in an informal dialogue. Check whether the target center you have perceived is indeed the appropriate one, and whether your actions are indeed yielding the desired outcomes.

Remember that you must remain adaptable. The triangle rotates to bring up different points as being paramount (on top) under different circumstances in the progress of your career. The shape of the triangle may change to accent differences in importance. Your position on this diagram depends upon what is outside, what is attacking, and how the triangle is being rotated, or shaped, by you.

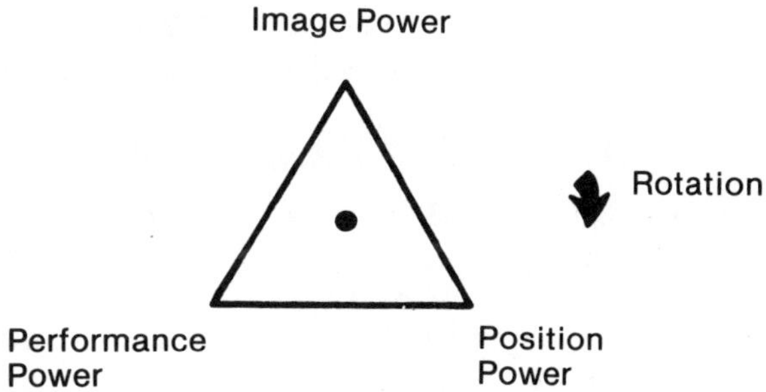

Figure 2.4

DETECTING THE NEED FOR ACTION

OVERVIEW A key difference between experienced and inexperienced managers may be illustrated by borrowing the definition of a problem from Kepner–Tregoe's *Rational Manager*. As shown in Fig. 2.5, performance starts at T_1 and runs to T_3 where it is expected to continue to point C at T_6. Instead something happens at T_2 or T_3 and by T_6, when performance should have been finished it is at point D. The problem is the deviation x—the difference between expected arrival at C and actual arrival at D, and the first manifestation was the change in slope during time T_3–T_4.

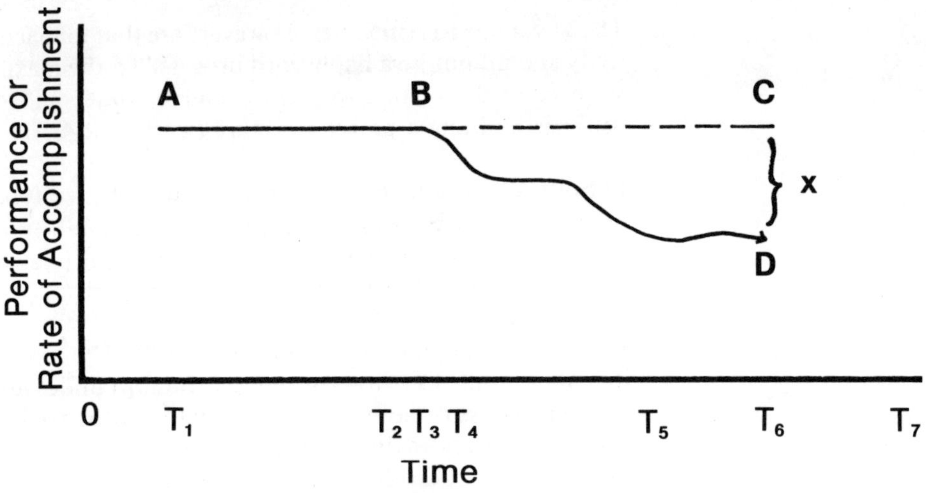

Figure 2.5

An inexperienced manager does not monitor the situation in a timely fashion and is "caught" at T_6. At best the manager checks with subordinates at T_5 when the deviation is too large to avert arrival at *D*.

The experienced manager will at:

(T_1) and beyond: Assess whether point *C* is within the capabilities of the organization and not just a theoretical dream. A manager must be aware that there are two types of planning involved in setting such targets: (1) outer directed; (2) inner directed. Outer-directed planning is done by people removed from those who execute the plans. Inner-directed planning is done by those who normally execute plans, but occasionally adopt a staff role to formulate them.

Since outer directed planners are remote from the daily group operations, coping with them can be difficult. The most frequent difficulty encountered is their setting *C* at ever increasing levels even where some stability, and group rebuilding, is preferable.

(T_2): Will detect a seemingly unrelated event as bearing potential impact on performance. Sometimes called a manager's intuition, close analysis reveals it to be:

- an in depth knowledge of the business of the organization and all forces which affect its operations and those of suppliers.
- a wide range of perceptive reading in a wide scope of newspapers, journals, information services, etc.
- dependable and widely connected circles of acquaintances in a wide range of endeavors. There is a spirit of mutual caring, trust, and information sharing built up with these people (the manager takes the initiative in calling and sharing information when it may be profitable to another even if no immediate gain to the caller is envisioned).

- an open and alert mind that easily and readily evaluates "how will this affect my business" when things happen.

(T_3): Will take corrective action to prevent deviations upon occurrence of an event threatening or holding the threat of hindering performance. The manager may even decide to ignore the threat and accept a deviation.

(T_4): Will have a better grasp of how to deal with deviations. Is correction worth while or does the organization have higher priorities? Can the organization be pushed to make the corrections?

(T_5): Will better evaluate the advisability of aborting the entire operation as opposed to further expenditure of resources in a vain hope for improvement.

(T_6): Can provide a better evaluation on how to use result *D* or whether it is worthwhile extending the operation to try to bring the result up to point *C*.

A control system involving appropriate time, monitoring, and action points makes it more likely a person will act in the experienced manager's mode.

TIMING

A manager is often reluctant to step in re the operations of subordinates. The feeling is that professionals who know what the objectives are should be able to resolve problems and attain objectives without further "meddling." If the manager who feels this way achieves satisfactory results in practice, then it is quite probably the appropriate style of management for those circumstances. However, it is also true that a technical group will be reluctant to express apprehensions or reveal technical difficulties "it will be resolving shortly." "Shortly" has a way of stretching until the project is delayed beyond the deadline.

An audit of the work should be continuous. An audit does not consist of second guessing the decisions reached or premature evaluation of the results. It consists of finding out *how* the group went about doing a particular project task, *what* information sources were used, *where* the project is going, *why* information is being sought or procedures are being performed, and *when* certain milestones will be achieved. Then ask yourself:

> Am I auditing a group activity or the actions of separate individuals? Do I want intramural competition (for independent approaches) or a focus on one promising approach? Which is actually being fostered?

Studying the answers to such questions can reveal far more about what is actually going on in a research, design, production, or other group than simply looking at "results to date."

Third Ear Listening. The manner of response to a question or the fact that a question is sidestepped may reveal more than the message content. This is sometimes called "third ear listening." The manager picks up two forms of clues: *verbal clues* (hesitancy, reluctance, inability to articulate, etc.); *body language clues* (the alert manager catches the glances that seek "permission to divulge" problems). Inability to express a problem or innovations to a manager to probe (asking permission to speak) may reveal deeper seated technical weaknesses.

Unlike the physical world wherein the measuring instrument itself changes the phenomenon under examination, the Hawthorne experiments revealed that observation in social phenomena has only transient effects. The "Observer Effect" may be used to provide the manager with guidance on timing intervention in the affairs of subordinates.

The observer effect tells us (see Fig. 2.6.):

$$T_{\Delta L} = T_3 - T_1 \propto 1/T_1$$

$$\Delta L = L_3 - L_1 \propto 1/T_1$$

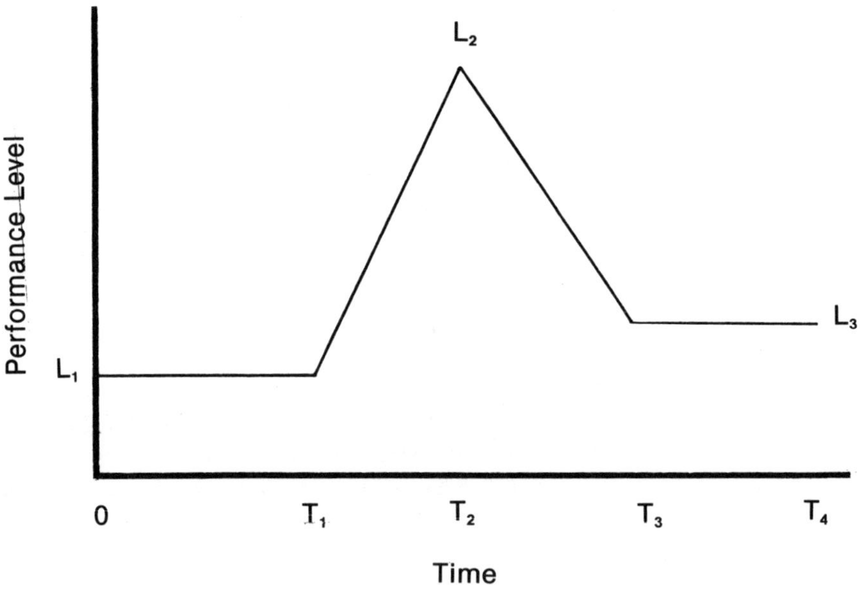

Figure 2.6

The length of time an organization will be changed (performance level increased) by the introduction of an observer is inversely proportional to the length of time the former (low) level of performance existed. The longer an organization has been allowed to exist at a low level of performance, the faster it will return to normal (or near normal) even if a change has been introduced (such as new rules) without a major "structural innovation."

The change in level of performance achieved by the introduction of an observer is inversely proportional to the length of time the former (low) level of performance has been allowed to exist. The longer an organization has been working at a low level of performance, the smaller the increase in performance that will be achieved by any change short of a major "structural innovation." If a poor practice has been going on in the office for a very long time, a supervisor who steps into the situation with cosmetic changes—new rules, policy, discussions, structures, etc.—will achieve a change in the organization to L_2 and be lulled into thinking something was accomplished. However, just as Hawthorne observers "melted into the background" and, after a short while no longer affected the situation, so will the changes just made melt into the background to no effect on the bad situation. The longer the objectionable practice has been tolerated, T_1, the shorter the time in which the change will hold, $T_{\Delta L}$ or $T_3 - T_1$, and the quicker the return to low or objectionable levels of performance.

Again, the longer the objectionable practice has continued, the less likely an appreciable change in level, ΔL, will be achieved by new rules and regulations (or their stricter enforcement) or even new control systems. L_3 will fall to L_1 regardless of transitory L_2 level if insufficient change has been made.

USE OF THE OBSERVER-EFFECT CONCEPT

The longer it is allowed to persist, the more difficult it is to change any practice and particularly objectionable practices. Tightening the screws or installing new control systems simply improves the level only over the period T_1 to T_3. Only if T_1 is short can L_2 be reached at T_2 and possibly held there by such methods. Confronted by a long T_1, a manager might try the controls approach to see if this can raise performance and hold it at L_3 well above L_1 and near L_2.

If the controls do not work, only a drastic change in the operation will have lasting effects. Reassign people out and bring new people into the group; rearrange the office and/or laboratory so that different individuals rub elbows; assign new groups to new group leaders; and then establish new controls.

One cannot interfere and correct every slight deviation from optimal performance, but it should be carefully watched to see if cumulative deviations over a long T_1 will result in a situation that will be difficult to correct.

BEHAVIOR

Assessing behavior requires a sound understanding of behavioral principles. Some people have great success in working with and supervising people based on their instincts; such people are the lucky few who have fallen into a situation in which their style is accepted or tolerated by others who mold their responses accordingly. For most of us instinctive management is not optimal—hence the need for study of leadership and its ramifications. It may be a trite observation, but it must be continuously enforced: Each personality is a separate and distinct dynamic factor within the work group.

It is not within the scope of this work to do more than briefly introduce the psychological/motivational theories filling the management literature—which we do under "Leadership" in the next chapter.

ASSESSING BEHAVIOR

Before they can be judged, people must have a clear idea of what is expected of them. This should be in writing and reinforced by appropriate discussions, interactions, and sanctions between manager and subordinate. See the material on MBO.

Understand that the complex interactions and expectations that are part of a working group can never really be captured on paper. However, they cannot be overlooked; the values, behavior, and attitudes which spell success are part of the working environment the manager builds through confidence in the group and, in turn, earning its confidence in him or her. The manager who treats people as professionals and sees that their work is on a professional level goes a long way toward building that confidence.

Do not depend on people learning "on their own" what is and what is not accepted behavior, methods of research, or how to investigate a problem. If they are learning from experienced and high performing peers, let them know that your expectations and appraisal of what they are learning is positive; reinforce it be reviewing it with them, praising their progress, and praising the "teachers."

If their behavior or work is not up to the standards you are setting, come right out and tell them so during your discussions, audits, and training sessions; it is a sad manager indeed who expects a subordinate to learn by implication when that subordinate is being dressed down for failure to meet a target.

Professional people are highly sensitive to the differences between professional and non-professional tasks. If the latter just cannot be shifted to more appropriate people such as technicians and secretarial backup, they should be eliminated where possible (cost/benefit analysis is appropriate here), minimized through changes in the way information is collected, as an example, and made less objectionable through such devices as preprinted forms, multiple uses for the same reports, etc.

Provide regular, unambiguous assessment feedback to subordinates even where the assessment is painful. A key way to avoid a "blame" and finger-pointing exercise is to focus on the "professional gap." The manager points out the results expected, the actual results achieved, and the gap between the two. Then discussion can better focus on how to close the gap—more help, more time, more technical training, more of what?—than on why it is there and who is "wrong." An interview focusing on gaps can make good use of nondirective interview techniques wherein the manager acts as a stimulant/mirror to the subordinate. After the preliminaries are out of the way and all involved are cognizant of the gap, the subordinate or the one responsible for the area in which the gap is a problem is free to explain it in terms of the cure and all other aspects appropriate to the discussion. The supervisor need only ask questions when the situation runs dry, need only express interest when a discussion needs reinforcement to continue and should reserve evaluative comments for last. The summary of the discussion and points for future action should be written up with copies for all concerned. Make this a learning session wherein assessment is used to channel behavior into proper modes.

Further Reference Topics

Critical Incident Listening or Third Ear Listening
Non Directive Interviews
Performance Appraisal
Books on Body Language

TALK TO YOUR SUBORDINATES

Maintain a consistent regular pattern of communications with subordinates to provide them with evidence that:

(a) Initiative, persistence, creativity, resourcefulness and results are recognized.

(b) They can be free and open to discuss technical, personal, or procedural hangups as they occur and can get your active help in their resolution.

(c) The manager is interested and "on top" of what is going on, and the interest is more than an image.

Maintain a routine of work assignment, work flow, and accomplishment of objectives that, while geared to individuals in the group, absorbs turnover in personnel and projects. Establish a system you can track and use for correction over the long run, not a single project or team of particular personalities.

BEHAVIOR CONTROL

Behavior control arises from:

(1) An appropriate style of leadership.
(2) A strong performance orientation in which effort is not confused with results.
(3) Reinforcement by recognition of accomplishment and emergency aid (intensive care) to lift substandard performance.
(4) Ruthlessness where substandard performance does not improve.

Comment—The Substandard Employee. Most people tend to think they are better than they really are; paper shufflers view themselves as engineers; shelf sitters and technical deficients view themselves as managers. They wait for the "big opportunity" to prove themselves. What they are really doing is expending effort instead of achieving results and accepting a strong effort as equal to a result.

A quick rule of thumb on accomplishment: If you as a manager owned the firm and the money for that effort was out of your pocket, would you be satisfied?

(5) Split low skill from high skill jobs and award to appropriate people. Separate low skill people from high skill people by carefully informing yourself of people's position and performance powers—not solely their image power. A high skill, high challenge job is usually a "reward" sought by high performers. They will resent a job going to someone less capable or less deserving.
(6) Make use of the position description as a tool. It should have objectives listed for that position, what routine duties are expected, the types and level of innovation in technical work expected, the type of projects which can be handled, and the career goals or path for that position.
(7) Focus on the resources subordinates require to do the job. They can provide some things for themselves by adapting different tools to do a multiple of tasks; other tools or resources they can do without by designing experiments around them. There are some only management can provide.

Comment—The Tools vs Results. It has happened a sufficient number of times to become almost a cliche, but awareness leads to appropriate response so the story is repeated here:

Jones company operated from rented quarters and electrical engineering had one oscilloscope, one old meter, and a host of "homemade" bread-boarded instruments. Jones engineers developed, in ten years, a mighty range of products.

Jones company moved to campuslike headquarters with concommittent upgrading for engineering. In ten years since, the number of new products or innovative features has been disappointing.

There are lots of reasons for this, but it has been noted that when expenditure of funds substitutes for ingenuity, the care and calibration of instruments substitutes for more demanding work, and purchasing's "depersonalization" substitutes for the talents of innovative people.

(8) Utilize the concept developed by the military, of "completed staff work."

(a) Work out all details completely.
(b) Consult with those impacted and those who may have worthwhile criticisms or comments to make.
(c) Study, write, restudy, rewrite.
(d) Advise the chief what to do. Do not ask him.
(e) Present a single coordinated action. Do not equivocate.
(f) Do not present long explanations or memoranda; correct soluions are usually recognizable.

(9) Utilize decision analysis in which the results withstand examination:

(a) How did the method affect the outcome?
(b) Why was this method chosen? Which and why were others discarded? The usual situation is the dearth of choices. As in cribbage, the discards are important. Management may know of reasons why a discard is better.
(c) Why was this outcome accepted? Which criteria were used and which/why were others discarded?
(d) What other outcomes were possible and why were they discarded?
(e) What changes in assumptions would have made a discard the chosen?
(f) How effective was the process (output achieved for input expended)?

(10) Get out of your office regularly and see for yourself what subordinates are actually doing. While there is no such thing as a suggested proportion of time to spend in this activity, it should never become deferable until your desk is cleared and the current crises are resolved. It should be a priority activity with some time allotted to it every day.

PROJECT CONTROL INTRODUCTION

Engineering proposals usually undergo several screenings before being assigned for accomplishment by a particular group. Then, all too often, they are "dumped" on the group manager with a list of specifications to meet, costs not to exceed, and due dates. At first glance it looks as if the project has been planned ad nauseum and the group manager has simply to put the project into work—copy the documents given to the group, figure out who does what, and assign accordingly. It is a rare occurrence that such a "system" will consistently show better than mediocre results.

The mistake was in accepting the screening and planning done on a high external (to the group) level for the planning which is needed within the group on its level. A manager needs a system for control of the work being assigned to be done by the resources under his/her authority.

The manager does not need elaborate scheduling tools; he does need information to schedule and track work under his jurisdiction and to see how his group's output fits into the scheduling of activities of which his group is a part.

A manager must establish objectively observable and measurable milestones for the work of his subordinates so that he has early warning of delays, slippages, and schedule problems.

STEP 1 PROJECT CONTROL

The manager should establish a series of "key indicators" which will reveal the "state of health" of all projects under way. If not already part of an MBO system, establish:

(1) Not more than seven macro key-indicators. Most people can readily assimilate and deal with seven variables at a time—a greater number of indicators may not, therefore, be comfortable for mental juggling.

(2) The key indicators can be backed up with computer printouts, budgets, etc., in endless numbers, but all the information in the backup should be encompassed by the seven macro-indicators chosen.

(3) Choosing the indicators is a tough task, but begin with a self dialogue and pencil and paper:

- (a) What outputs do my superiors watch? What measures do they use?
- (b) What measures can I derive from the above?
- (c) What measures or items have I been examining, consciously or unconsciously, which are tipoffs something is wrong?
- (d) If I could wish for at least three indicators my bosses would use to make this group look good, which would I choose?

(e) If I wanted, to project an image that this is the best managed group in the firm, what physical activities would I want the top brass to see on a surprise inspection? Can I measure these?

(f) I have seen things in other groups, in and out of the firm, which gave me the impression of a mediocre operation. What are they? Can I measure them?

(g) Some of my subordinates are superperformers—a piece of their work coming into my office has an elegance and image that stamps its origin immediately. What are the items I observe, hear, smell, and feel? Can I measure them?

(h) Some of my performers are quite poor. Can I measure the outward manifestations of their work?

(i) Examine the list to see which measures are simply and easily collected. Simplify or discard those which call for elaborate data or expenditure of large amounts of time in collection.

(j) Combine, weed out, recombine, rethink the indicators. Would you be happy to have your performance judged on the measures remaining? Weed out those which would be affected by matters beyond your control.

(k) Establish a framework for tracking the measures left. Those which cannot be done by paper and pencil under a few minutes per day (at most) are discarded.

(l) Set up the tracking system for the remainder measures. Keep the ones which turn down fastest when things start to go wrong, which make the group look good when it deserves to look good, and which can be dissociated into more detailed measures when necessary.

(m) Operate the group using the measures to collect baseline data. Check with the literature for additional measures in your field, improvement of the ones you are using, or for comparison data.

(n) Swap information with others in your position at conventions, etc.

(o) Get approval and recognition from your superiors and subordinates for the measures—modified if necesary—and use them.

(p) Provide your subordinates with regular feedback on how the measures look.

EXAMPLES

(1) Number of items late—milestones missed.

(2) Number of complaints, revisions, or bounce-backs.

(3) Results you mentally classify (number of each):

A • innovative/super performing

B • average acceptable

C • subpar—think some more.

(4) Completed projects or assignments—number.

(5) Points of transfer between departments or workers. (When something is handed over to someone else, how does the receiver feel? Which end of the stick did he get?) Sometimes a person will get everything shipshape before handing it over and make the work of others much more efficient; at other times, an over-emphasis on getting everything "just right" may have negative consequences for the operation as a whole. We are reminded of the homemaker who polishes all the silverware, before handing it over to the cleaning lady, to preserve a certain image.

STEP 2 PROJECT CONTROL

Each project, assignment, or work request should undergo the manager's planning step before release to the group for work. Along or, preferably, in concert with those most likely to be selected for the assignments, screen the managerial aspects of the project as carefully as the technical aspects.

(a) Determine what will be the scarcest resource—time or money or talent or?
(b) Determine the areas of greatest
- Uncertainty
- Challenge
- Opportunity

(c) Determine the tightest constraints.
(d) Establish the technical and managerial objectives. The technical objectives may be the specifications to which the work must conform. The managerial objectives reflect the accomplishment (outcome) and process. Statements should reflect the priority this job has vis-a-vis others in house, this job vis-a-vis "emergency" requests, the space, time, talent, etc., resources. Outcome reflects results your firm wants. Process reflects efforts.

EXAMPLE OBJECTIVES FOR STEP 2

Technical

To design and build prototype minicomputer to meet specifications A-1, 2, 3, by July x, 19xx, at a cost of $100,000.

Managerial

Outcome

To design and build prototype minicomputer to meet specifications A-1, 2, 3 by July x, 19xx by utilizing in-house engineering resources, present facilities to be enhanced as determined by events in plan "Z," and materials from known vendors A, B, C, on negotiated basis within cost envelope of $100,000.

Process

(1) To have engineering trained to do job through after-hours seminars at local college by January x, 19xx, at a cost of tuition reimbursement for 10 people.

(2) To have a preliminary list of supporting facilities needed by January xx, 19xx at cost of 100 man hours engineering time.

(3) To accord this project highest priority of all jobs in house. All requests for diversion of engineering talent or support equipment away from this job are to be cleared at company level X with no diversion allowed more than once per month for 3 days each time max. (**Note.** Such a situation may well lie in the realm of fantasy. In which case, try to get management to agree that the time and dollar constraints will be incrementally adjusted for each ''violation'' of priority.)

STEP 3 PROJECT CONTROL

Choose a method for keeping track of the sequence of milestones to be reached on the plan and the actual accomplishments. These methods need not be computer driven complex management information systems. Rather, they should be simple, fast, and amenable to your personally updating them with only pencil and paper; all other complex systems run by the firm are additional aids for you, but not primary to internal management of the group.

Updating the indicators is a job not to be delegated. The manager must go through the mental exercise of:

(a) stopping all other duties at least once in the week;
(b) examining closely what has been done in the group;
(c) judging what has been done—an accomplishment worth noting or simply lots of activity;
(d) noting the judgment made on the indicator sheets;
(e) taking follow-up action.

The system to be used has the following managerial features:

(a) It is integrated. It keeps watch of the several indicators chosen simultaneously, not only one or two at a time.
(b) It encompasses the traditional statistical quantity and quality accomplishment measures along with the others deemed necessary.
(c) It encompasses some measures of organizational relationships—work flow, relationships with superiors, clients, peers, the matrix setup of a project organization, etc.
(d) It has indicators which change slope to reveal breaks in continuity of work flow, even where such breaks might otherwise be unobjectionable,

such as creeping lateness of routine reports, creeping increases of slight deviations from plan, boredom, complacency, and where actual accomplishments are ignored.

(e) It has broad outlines prepared of the actions you will initiate and the point at which they will be initiated—a threshold response level—at the rise of symptoms of trouble.

PERT/CPM Program Evaluation and Review Technique (PERT) is also pushed, in slightly different format, under the Critical Path Method (CPM). In the former system, the accent is on keying the network diagram to the appropriate activity end point (event). In the latter system, the focus is on the activity path. The distinctions are purely technical as a manager might most profitably use a combination of the two.

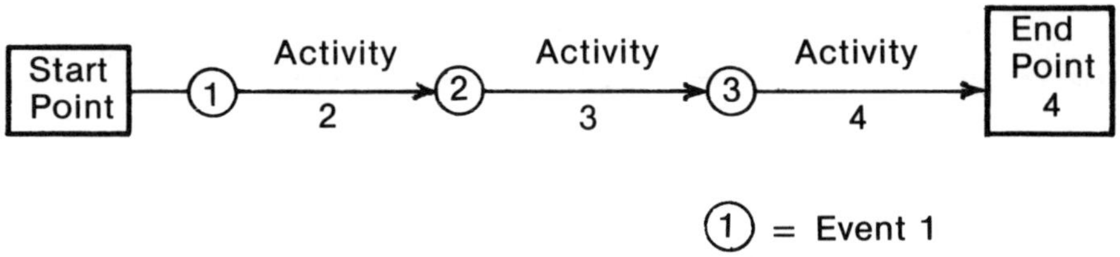

Figure 2.7

The analysis is done as follows (see Fig. 2.7):

(1) Lay out the end-point objective.
(2) Lay out the milestones your group must meet with specific accomplishments for each.
(3) Sequence the accomplishments—which depends on which and the timing of the accomplishments. Physically this may be done by writing the accomplishments on 3 × 5 index cards and laying them out from first to last from left to right. Those items amenable to parallel accomplishments are parallel to each other on the card layout.
(4) Connect arrows from item to item showing the appropriate paths of completion. Label these arrows to indicate the appropriate activities. Note times on the arrows to be allowed for the activity and dates in the event circles or squares for completion times.

A complete analysis would now require the calculation of the longest path in the network from start to finish—the critical path. However, as a management tool here, the inclusion of a series of dates on the diagram effectively prevents a pileup of activities beyond the due dates. Refer to PERT/CPM texts for methods of calculating critical paths. Here, the tool is used as a visual tracking device only. The information on the 3 × 5 cards and all associated arrows need only be copied on a single planning sheet with careful monitoring to assure actual progress equals planned progress.

PERT/CPM EXAMPLE Design and build minicomputer to meet spec. A–1, 2, 3, by July x, 19xx at a cost of $100,000.

List of Milestones: Each box represents completion of activity (end point of activity).

Prelim. research	1	Tie-in circuits	6
Layout of Unit	2	Programming	7
Power supply	3	Documents	8
Logic circuits	4	Test & evaluation	9
Cabinet & controls	5	Completion	10

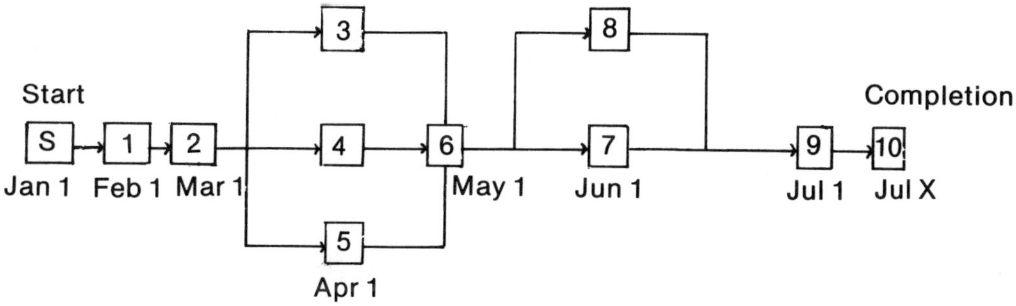

Figure 2.8

Each activity arrow in Fig. 2.8 could be labeled with the resources to be used, the names of people assigned to do the jobs, the facilities involved, and budget dollars allocated.

Note. The dollar budget should be carefully integrated with the time budget. Time budgets, or time schedules, incorporate management decisions on definite blocks of time and dollars to be allocated to specific tasks; thus, even though the time schedule is the product of PERT or CPM analysis, it is never divorced from dollars.

Dates are assigned for completion of activities. With this on paper, it may soon develop that waiting for the results on a monthly basis leads to schedule problems when one or more run late. Should it be evident there is a potential problem, isolate the segment of the diagram which represents the time span or activity involved. Figure 2.9 shows that the Jan 1–Feb 1 segment might be a problem. A "blow-up" of this segment shows activities in greater detail and milestones in weeks instead of months.

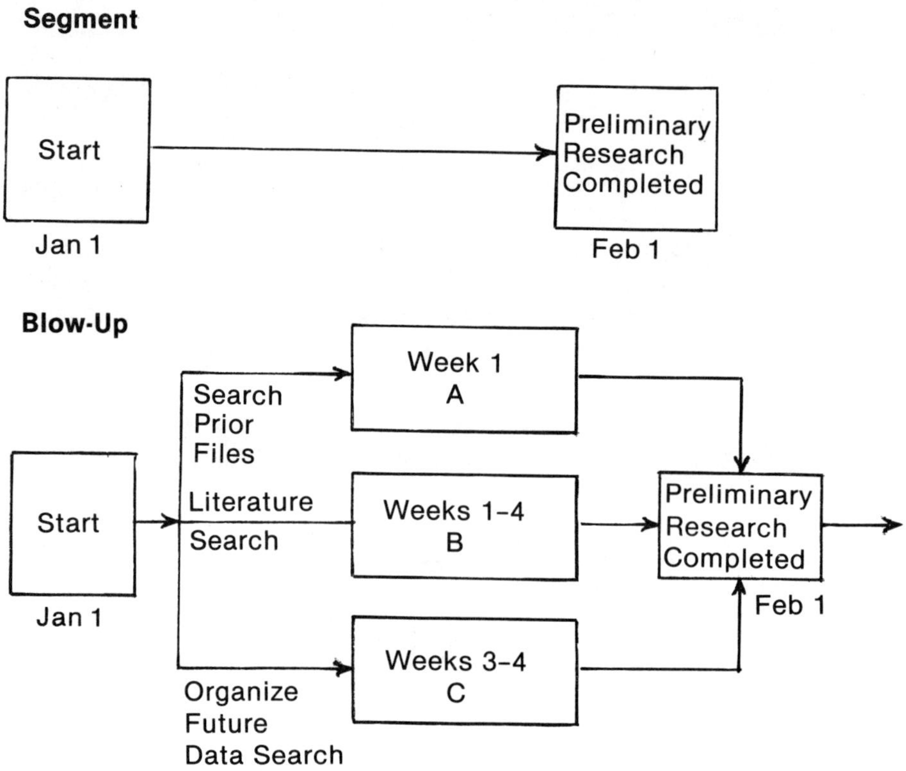

Figure 2.9

The manager is the judge as to whether a task has or has not been completed; avoid reducing your aspirations to meet the level of accomplishment where better work could be done. A performance oriented manager can quickly audit most situations and render praise where deserved and vice versa.

Caution: If the diagram becomes the measuring device, aspiration will be reduced to the lowest acceptable level of commitment.

PROJECT CONTROL —FINAL COMMENT

A very simple system for pencil and paper tracking plus a coherent set of indicators for the group's work will help you run a *performance based* operation.

CHAPTER 3
Leadership: Three Approaches

Chapter 3
Leadership: Three Approaches

BASIC CONCEPTS

The new manager frequently "falls into" a style of management based more on feelings and reactions to events than on a careful analysis of what might be best in a given situation. The result can be a blend of features picked up consciously and unconsciously from previous bosses, teachers, parents, and others. At best, this approach may lack consistency; at worst, the amalgam may be totally inappropriate to meet changing circumstances in managing a group.

Where does the new manager turn if he or she decides to investigate styles of leadership and to adapt traits to insure success? A major source, of course, is the literature which discusses leadership in three ways. The first type of discussion looks at leadership as an overall pattern of behavior and attempts to derive theories for such patterns; the theory accepted, success patterns can be established for the manager to follow. We will briefly review four such schools of thought: (1) Rensis Likert; (2) Blake–Mouton; (3) Douglas McGregor; (4) Victor A. Thompson. There are dozens of other names we could have included here, but these four probably best exemplify well-known applicable theories.

Leadership is also discussed as a process. Here the literature discusses the processes of motivation, communication, control, planning, and a litany of duties. Books are written on "management by objectives," "management by exception," "participative management," and others. The process school of thought was probably best exemplified by Henri Fayol (who wrote of the Plan, Organize, Command, Coordinate, Control concepts in 1916) and is logically extended to work in MBO by such men as George Morrisey. We will briefly review the work of these two men and touch on the aforementioned processes as well.

Finally, the third school of management examines leadership traits. Researchers have studied outstanding leaders and attempted to discern just what particular or distinguishing abilities, characteristics, or distinctions they have. This school of thought is probably the easiest for a new manager to assimilate because it permits careful, conscious, self-development. We have surveyed publications ranging from *The Prince* to psychological studies. We cover here a short list of traits we feel most practical for self-development. Because traits are derived from case studies and present a body of thought most amenable to self-development, we label it "the empirical school."

WHY WE NEED THEORY

In this book we stress a "how to do it," "action oriented" approach. Why then, this diversion into theory?

Because the *theory* is part of the *doing*. Experts who have studied the operation of the managerial mind tell us that its operation takes place in a series of steps (not always in sequence, not necessarily all steps present in any one situation). Simplified, their theories tell us that the first step is that of "becoming aware" or recognizing the existence of a problem or an opening to exercise initiative. Following the awareness stage, the mind defines the situation, searches for information, analyzes (or solves) the problem, evaluates the intangibles, and executes an action.

The trigger, the item which "sets off the mind," is the manager's awareness or recognition of something stimulating his senses. This key step is facilitated by the mind that has stored in it at least some theory. Those who claim to manage by instinct or those who claim to manage without theory, have, in reality, developed a theory or sets of theories over their lifetimes and use these unconsciously to trigger their thinking processes. Some of these "life" theories are good ones: the person has been exposed to good influences, internalized excellent learning experiences, and developed a good measure of insight into this accumulated knowledge. The converse, of course, is equally true. Or, a person may have developed no relevant theories and no insights; the situation is totally foreign to him. "Dropped" into the manager's position, such a person is like the unprepared businessman dealing with a strange culture: he says the wrong things, offends when he does not mean to, misses opportunities for initiatives, and attains mediocre results. Worst of all, he has no mental framework on which to "hang" his experiences so that learning is not cumulative and mistakes are repeated.

Awareness is a "mind state" characterized by:

1. Openness to experiences. Stimuli are freely relayed through the nervous system to the brain where they latch onto already present ideas or insights. They confirm these ideas, create new ideas, or confront them causing a recasting of the old and new in different ways. Theory implants the ideas which increase receptivity and provide the material for recasting.

2. The ability to perceive things as they are. Instead of perceiving all things in predetermined limited categories, the individual can deal with experiences falling outside his usual range of categories. The broader the theoretical background, the more easily the mind assimilates different ideas and deals with them constructively. Every problem is not automatically dealt with in yesterday's terms, or with a fixed reference (such as the "technical" approach), that may yield a poor result.

3. A turning up of the mind's "sensitivity" control. The ears become more receptive, the eyes more keen, so that any stimulus from the environment enters the brain from more than one set of sensory nerves. At the same time, the flow of messages within the brain is enhanced by the flow from prestored concepts. Theory gives the brain the material to generate new concepts and to apply them on the job.

Theory helps the mind in the "definition" of problems, decision situations, or opportunities for initiative. The mind may recognize something—an amorphous situation. It then attempts to put the situation into words, pictures, sounds, smells, or even tastes so that it can tell itself what it is that has to be done next. This structuring or shaping of an amorphous situation is known as the definition process. A manager will use words to define most management situations. In this sense, management is more limited than engineering which has the luxury of using diagrams, graphs, flow charts, symbols, and models in addition to words (but too many engineers become inarticulate when deprived of these aids). The vocabulary of a person with a broad theoretical background contains many more words, for use in many more contexts, than the person who depends solely on life experiences. Hence, definitions come to mind much more readily in the person who has studied theory.

Theory helps in gathering facts and impressionistic information. It points out what to look for, what to expect, and, quite possibly, why. The information search may reveal a solution or the futility of a particular direction of the work. It provides the structure for assimilating experiences in a coherent and logical fashion.

Finally, theory also helps, in obvious ways, in the analysis, evaluation, and implementation stages.

The theory review presented here is designed to take the manager from "ground zero" to a position of some familiarity with the body of managerial knowledge in the textbooks. It is material fundamental to the mental processes of the new manager and provides an appropriate stepping-stone to further study.

LEADERSHIP AS A BEHAVIOR PATTERN

THEORY: RENSIS LIKERT

Styles of Leadership[3]

A manager's style can be classified by a location on a continuum which has at one extreme (call it the left side) a centralized authority, exploitive, untrusting, sole control leader. At the other (right side) extreme, is the leader by full participation. This leader uses groups for decision-making, builds supportive relationships with subordinates, and uses subordinate peer relationships, group goals, group norms, etc., for group supervision.

Moving from the left extreme toward the right, the first intermediate point is the Benevolent–Authoritative style of leadership. This is a benevolent master–servant relationship, a patronizing and sometimes paternal system. The manager gives subordinates the impression that service in my group is an "honor." "Work well in my group and I will see that the best is done for you." The manager places little trust in subordinates.

The second intermediate point is Consultative Management. Here the manager has a great deal of trust in subordinates, but that trust is mitigated by the fact that "since I am the boss, I have to keep a hand in everything." In this type of atmosphere, the informal organization flourishes and constantly fights for more final say against the limits the manager establishes in the various areas. Such conflicts can become intense: the group sees them as a fight for more trust in the form of having a greater say in affairs of the organization. The manager sees them as an encroachment beyond tolerable limits on the right of the organization to manage. At the extreme right side of the spectrum is Participative Management. Under the participative

[3]*Likert, R.,* New Patterns of Management, *New York: McGraw-Hill, 1961.*

management style, the formal and informal organizations are the same. Since the group has so much control, there is no incentive for it to establish an informal organization; informal groups which do establish themselves are usually short lived (susceptible to take-over by a group member who disagrees with the approach and believes that he should be the take-charge guy who can lead the others to bigger and better things—the Weimar Republic was succeeded by Hitler).

COMMENT: REALITY

Managers of engineering organizations may start out liking the idea of participative management but quickly run up against the following types of constraints:

(a) The manager may have more information than the group for making a decision. Time constraints or confidentiality barriers may prevent information dissemination.
(b) The manager feels the blame will fall on him or her; why, then, let the group come up with a decision?
(c) If the group comes up with a decision the manager dislikes, what then?
(d) How does one handle strong parochial interests or unpleasant decisions the group avoids?

The manager quickly slides to consultative management wherein the group suggests but the manager decides. This is a good style as long as the group gets its way. But it backfires when a string of group recommendations is rejected for whatever reasons the manager may hold. Then the informal organization becomes an active opponent unless the group can be satisfied that its inputs were truly considered and that the ''consulting'' was not fake.

The benevolent style patronizes subordinates and is quickly seen as a mask shielding the master from the servant in a supposedly democratic world.

The authoritative style falls of its own weight when the manager has to make decisions on matters in which subordinates have the information. It is accelerated by the authoritarian's tendency to decapitate messengers bringing bad news; soon he gets no messages—only echoes.

USE OF LIKERT'S THEORIES

Most managers do not adopt any one style on Likert's continuum. They switch from one to another to avoid complications. The best use of Likert's ideas may well lie in ex post facto analysis of why a group's operations were not successful.

THEORY: BLAKE-MOUTON

The "Managerial Grid® "[4]

The vertical axis of the grid established by Robert Blake and Jane Mouton represents 1 through 9 "degrees of concern for people." The horizontal axis represents 1 through 9 "degrees of concern for production." The lowest degree is a minimal concern and vice versa. Blake–Mouton identify five major styles of management on the grid as given in Fig. 3.1.

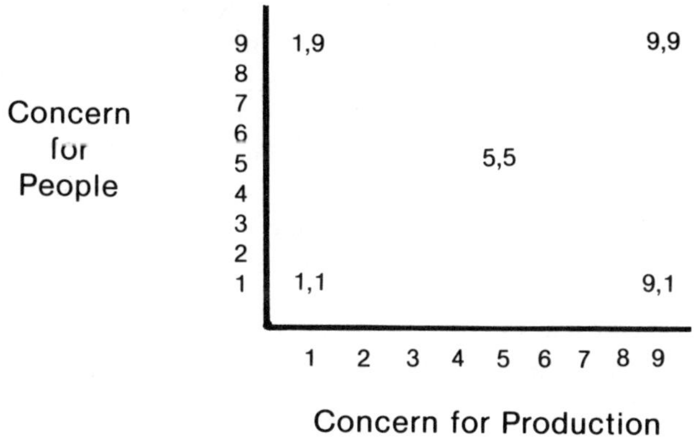

Figure 3.1

(9,1) *Task Manager:* Nine degrees of concern for production and one degree for people. People are viewed as tools of production and the manager is interested only in getting the job done.

(1,9) *Country Club Manager:* Harmony breeds contentment and contentment yields production—if the practitioners of this style of management are correct. Runs parallel to the feeling about contented cows.

(1,1) *Impoverished Manager:* Fear of rocking the boat (making waves) yields a manager who does the absolute least to get by. Bureaucracies with "protected positions" working toward intangible or difficult to measure results seem to be a breeding ground for this style.

(5,5) *Pendulum Management:* Pushes as events require for production or for maintenance of morale. Works hard at smoothing things over to keep up one or the other when problems arise.

[4] *Blake, R. R., and Mouton, J. S.,* The Managerial Grid, *Houston: Gulf Publishing, 1964.*

(9,9) *Integrative or Task/Team Manager:* Continuously examines the operations of the organization to see that it is meeting performance goals while at the same time building the production team. Delegates responsibility for results, gives information freely, and builds team commitment to achieving the group's goals.

Blake–Mouton have found that most managers have two styles of management. The first is a dominant style—one which is used most often and the second which is a backup style. The backup style comes into play when the dominant or preferred style does not seem to work. Styles need not be at the extreme points illustrated. In practice, people attending grid seminars fill out extensive questionnaires which are graded; managers may have 7,3 or 8,2 dominant styles with 9,1 or 9,2 backup styles. The style may also change with the situations confronted or the environment set by management. In fact, a great deal of trouble will be encountered by a manager who does not read correctly the style of management desired by the "top." The boss may want production regardless of its cost to the team while the manager may wish to allow some slippage to avoid coming on like a slave driver. Many managers also express a preference for a "modern" (i.e., participative) management style, but subconsciously believe it to be a luxury which they will employ "as soon as things improve a little more." In other words, they'll use the style when they "can afford it." Meanwhile, Marine Drill Instructor!

USE OF GRID

The best way to use the grid, without having access to its extensive seminar material, may be for the manager to conceptually analyze the organization. Does the organization hold the view that there is indeed a tradeoff: you can't get profits without willingness to sacrifice people? The all too-quick breakup of successful engineering teams the moment a particular contract is lost or reluctance to allow team development through seminars or convention attendance show deemphasis of the "people concern" axis. This firm wants to get today's contract done today; tomorrow's contract can be staffed from the colleges if necessary. The manager moving upward in this company is not appreciated for anything but the degree of accomplishment re the contracts held by the firm (known to cynics as "what have you done for me lately?"). He or she may cue his/her behavior and resource expenditures (particularly time) accordingly. In short, analyze the firm's real emphasis, its position on the grid, and establish yours accordingly.

THEORY: DOUGLAS McGREGOR

Theory X–Theory Y[5]

McGregor analyzes managers by examining the assumptions these managers make about the people working for them. These assumptions may generally be classified under either of two contrasting theories. The managers adopt a style and deal with people predicated on their belief in one or the other of the two theories (or their underlying assumptions).

Theory X

(1) Average human dislikes work and will avoid it when possible.
(2) Because of this dislike, people must be coerced, threatened, controlled, punished to get the job done.
(3) Average human prefers direction, avoids responsibility, has little ambition, and wants security above all.

Theory Y

(1) The expenditure of physical and mental effort in work is as natural as play or rest.
(2) Man will exercise self control and effort to meet objectives to which he is committed.
(3) Commitment to objectives is a function of the rewards associated with achievement.
(4) The capacity to exercise a high degree of imagination and creativity is widely distributed in the population, but within modern industrial life these potentialities are suppressed and therefore only partially utilized.

USE OF X-Y THEORY

A manager's style is strongly influenced by the basic assumptions he holds about people. However, the corollary also holds true: treat the people in the theory X fashion and they will respond in the theory X manner. This gives a false sense of success to the manager who then brags about "knowing what motivates the work force."

A more subtle paradox appears when the manager talks theory Y, sets up Y-type procedures, but is perceived by subordinates to believe in theory X. They respond as X, thereby confirming the manager's suppressed suspicions that X is the correct theory.

Knowledge of the theory provides a framework for examining your relationships with subordinates. If they are acting as theory X people, is it because they truly are X, or is it because they perceive you accordingly? In dealing with professionals it is generally a more successful strategy to assume and act on a theory Y basis.

[5] *McGregor, D. M.,* The Human Side of Enterprise, *New York: McGraw-Hill, 1960.*

Theory X–Theory Y Management Style Self Test

Check the degree to which you agree with the statement.

	Strongly Agree	Agree	Disagree	Strongly Disagree
Budgets and schedules are paramount: push subordinates as hard as necessary to meet them.	☐	☐	☐	☐
A good control system must catch the slightest deviations.	☐	☐	☐	☐
It is up to the manager to plan the group's work and the work of every individual in it.	☐	☐	☐	☐
The manager sets the group's goals and objectives and it is the responsibility of all subordinates to meet them.	☐	☐	☐	☐
A good manager spends the bulk of his or her time reviewing subordinates' work and ironing out their problems.	☐	☐	☐	☐
A manager gets paid to make the decisions.	☐	☐	☐	☐
A manager who prevents deviations from plan will prevent most unpleasant surprises.	☐	☐	☐	☐
Employees who oppose management's decisions are team wreckers and create problems they should "bury" before they interfere with operations.	☐	☐	☐	☐
Most personnel problems in an organization could be solved by raising pay scales to such levels the firm can attract and need retain only top notch people.	☐	☐	☐	☐
An employee who makes unauthorized decisions should be fired after the first warning.	☐	☐	☐	☐
Every so often, some "heads should roll" to keep the rest of the employees on their toes and show them that management does not tolerate anything but top performance.	☐	☐	☐	☐

SCORING AND RATIONALE

Give yourself a +2 for every item with which you strongly agree.
Give yourself a +1 for every item with which you agree.
Give yourself a −1 for every item with which you disagree.
Give yourself a −2 for every item with which you strongly disagree.

Total your score and enter it on the scale below:

−22 Theory Y	0	+22 Theory X

The rationale behind the answers is as follows:

Employees do not need to be "caught" in every deviation and neither do you need a control system to catch such items unless you do not trust the employee (theory X) to work without such detailed supervision.

The love of budgets and time schedules is dictated by the demands of the hierarchy. Acceptance of these limitations without consideration of mitigating circumstances, or the possibility of coming up with a better product in exchange for some relaxation, means that you are in favor of absolute enforcement of the limits (theory X). An absolute approach without reservation means that people cannot be trusted and must be forced to "toe the line."

Team wreckers are not always wrong. Sometimes the "devil's advocate" role must be allowed to prevail because it can point out and prevent serious mistakes. A manager who suppresses dissent and feels it necessary to bury the problems dissent creates does not trust his personnel (theory X).

Even the best pay scales in the world and even the best and the brightest will still create personnel problems for any organization. People have moods, have changing expectations, and, in particular, once their needs for good salaries have been met, will experience needs on a higher plane and make demands on the organization. Theory X managers too often have the feeling that a heavy carrot and a big stick are all that are needed to deal with employees.

As you examine the statements, notice that if you agree with them, you are somewhat mistrustful of subordinates, feel that they need careful supervision, careful layout of their work, and restraint of opposition to your policies. This is a Theory X manager.

THEORY: VICTOR A. THOMPSON

The Hierarchy and the Specialist[6]

Superimposed upon the modern organization's highly elaborated division of work is also a highly elaborated hierarchy of authority. Such a form of organization is called a bureaucracy. As this form of organization has developed, the most stubborn problem it has is securing cooperation from and among individual specialists. Organizations spend much time and effort forging a blend of skills among specialists to achieve goals. Being cogs in such a bureaucratic machine causes people to lose control of their lives—they have a sense of powerlessness, of alienation, and respond accordingly by manipulating the organization, conforming to it, or dropping out.

Thompson's analyses revolve around relationships between specialist and hierarchical roles. At the upper levels, the executive job and hierarchical role become synonymous, but at the lower levels the gap between specialist and hierarchy grows wider each day. There then develops a growing gap between the right to decide, which is authority, and the power to do, which is specialized ability. This gap grows because technological change occurs at a rate faster than changes in cultural definitions of hierarchical roles. This produces tensions; willingness to cooperate is strained.

There is, in modern organizations, a growing imbalance between ability to do and authority to decide. Specialization leads to problems with the hierarchy. Higher status persons become dependent on lower status persons; it yields the same frustration suffered by a highly trained electrical engineer who must deal with the local electrician when the lights go out.

The boss–subordinate relationship is formally unilateral with rights running in one direction from the boss to the subordinate. The advance of technical knowledge and specialization converts the relationship informally into a unilateral one with ability running from subordinate to boss. Whereas the boss retains full rights to make all decisions, he has less and less the ability to do so because of the speed of change in science and technology. This reversal of relationships may be followed by a partial role reversal as well: the specialist makes more of the decisions and the boss must become more involved in the work. This is a distortion of cultural and formal authority patterns with the end results being strain on the status system and potential conflict.

The negative effects of the reversal may be reduced by the manager who is aware of what is going on if he or she recognizes that the authority of knowledge has earned the technologist a measure of authority in a decision

[6]*Thompson, V. A.,* Modern Organization, *New York: Knopf, 1961.*

situation. The manager will then seek to generate the recommendations from the technologist and reserve personally the decision (selection) only where necessary and unavoidable; otherwise, participative management can be practiced. Insecurity and frictions give rise to behavior patterns that are counter productive to the organization's achievement: aloofness; personal insecurity; power grabs; exaggerated dependence on regulations, procedures, and standards for control. This is followed by a dampening of innovation or any conduct out of line with generally accepted patterns. "Don't rock the boat."

APPLICATION When a superior must come to a subordinate for information, there is an immediate role reversal. The subordinate has control over the situation in the quantity, quality, timing, and format of the information provided while the superior is, in effect, a supplicant. The psychic costs to many superiors are so great they avoid face-to-face requests by hiding behind memos. In other cases, where the superior can easily deal with this reversal, the subordinate then exacts a huge psychic cost by a patronizing, or disrespectful attitude or plain ineptness in appreciating just what the superior needs from that transaction.

Engineers, in particular, are prone to poison their relationships with superiors whenever the latter need something from them. The technical specialist will "rub in" the fact that the boss has fallen behind the state of the art. The engineer will: (a) provide information grudgingly—an intrusion into his "more important" technical duties; (b) be disrespectful in word or gesture when carrying out this request; (c) take little or no care in ascertaining what the superior really needs and provides too little, too much, or the wrong response; (d) will take little or no care in pointing out the significant from the trivial; (e) lose sight of the fact that he/she and superior are in the organization to achieve the same objectives. Since the supervisor has to suffer psychic costs in dealing with the role reversal, any of the sore points magnifies the difficulties leading to overt or covert conflict instead of mutual trust.

Role reversal damage is minimized when a manager believes in and practices participative management: he or she recognizes the authority of knowledge of the technologist. It is also minimized when the manager sees himself in the role of facilitator, catalyst, coordinator, allocator of functions, task allocator, and manages by objectives as opposed to the usual concept of command equals control or vice versa.

Role reversal damage may also be minimized by recognizing psychological hang-ups in subordinates. There are those who may be withholding the required information because of their need to feel indispensable. A person who transfers vital information to a superior (or to a peer, or even a subordinate) may feel that his reason for being, his unique value to the organization or project, will disappear. Often this hang-up is detectable in the person who never writes anything down, but must always be consulted in person.

Finally, the alert manager will deal with his/her superiors on the basis of enlightened psychic cost minimization practices. The manager will build trust, confidence, and a reputation for integrity by providing superiors with full information, in proper format for their needs, and in a respectful, professional manner.

USE SUGGESTIONS

The first step in dealing with the hostility is to become aware of it. You have read the preceding material. Now, look and listen during your interactions with others. As you see hostility develop, start, consciously, thinking of ways to deal with it—immediately!

The second step, in dealing with this hostility, is to defuse it. This can be done in either of two ways:

(1) *Seeking Assistance Method.* The police department of the City of New York developed ways in which an officer should seek information and help during an investigation. The department's research revealed that the format of the request determined, to a great degree, the success or failure the officer achieved in obtaining what he needed. The successful format of a request for assistance looked like this:

a. Identify yourself.
b. State the reason for your inquiry.
c. Ask, in specific terms, for the specific help you need.

If you are a stranger to the individual you are dealing with, or suspect that the individual just might not remember which department you are from in the firm, be sure not to leave out item (a). Don't consider it redundant—try it even in your after hours dealing with tradesmen and others and watch the reaction as you become a "somebody" instead of remaining a total stranger.

When dealing with a subordinate or someone with whom this step is, truly, superfluous, it might be a good idea to replace it with an "empathy statement." This is followed by a statement which channels the conversation into your area of interest. Then proceed to item (b).

Item (b), which gives the listener some of the background to the request you will shortly be making, should get the listener interested. It should help motivate the listener to supply the needed information because he or she is now a "working part" of the information stream. It is a professional approach because it is informative along with being explanatory.

Item (c) is a specific request for the specific information needed. It may often include an additional request for "whatever else you think I should know on this matter," but don't leave out words that directly ask for assistance. Too often, a manager will walk away from a subordinate unhappy with the interchange that just took place. The subordinate remains with a question in his mind as to just what the boss really wanted while the boss has the feeling that he did not get what he needed. Analysis of the interchange most often reveals that the manager never actually asked for the item. He may have gone through parts (a) and (b), but left it to the subordinate to "infer" that the boss needed item x or item y.

Example statements of seeking assistance:

> "You seem to be awfully busy here on projects Y and Z, and I hate to interrupt."
>
> *(Empathy statement)*
>
> "But I have this matter of project X that is pressing."
>
> *(Channeling into the area of interest)*
>
> "Project X is falling behind and top brass is looking at some minute figures on where we spent our time."
>
> *(Reason for the inquiry)*
>
> "Could you please give me a hand in digging out the time numbers and relevant figures for the project—we want to look at items p, q, and r."
>
> *(Request for assistance)*

(2) *The "request-to-reveal" method.* Here, the manager comes right out and asks the subordinate about the difficulty—but in a very special way. The manager "owns up" to his own feelings and then invites a response.

> "Joe, I seem to have some difficulty with our communications; I'm confused and a bit uncomfortable."
>
> *(Owning up statement)*

"I may have been rushed, and not as informative as I should have been, about what I needed the information for, but I would like to see if we can get back on the beam."

(Reason for owning up and channels conversation into areas of interest)

"Do you feel it too?" "Is there something about these requests that I can clear up?"
(Note that the request is to discuss the communication and interaction —not the technical content of the material)

The subordinate may not want to talk about the problem that is developing, but it still is a good opening for the boss to be willing to discuss the communication process instead of "burying" it under the technical content. It shows a concern for the feelings of the subordinate and airs the difficulty inherent in the role reversal.

LEADERSHIP AS A PROCESS

THEORY: HENRI FAYOL

The Process of Management[7]

While the management process has been studied over hundreds of years, a commonly accepted landmark work focusing on the "process" per se appeared as a special issue of a 1916 French mining journal. A man who had taken a near bankrupt mining firm and built it to number two in size among all French firms examined management and leadership from the top down.

The elements of management are:

(1) *Plan:* To envision the result, the line of action to be followed, the stages to go through, the methods to use.
(2) *Organize:* To provide a business with everything useful to its functioning—the 3 M's: Money, Men, Materials.
(3) *Command:* The organization, having been formed must be set going; it is the giving of the orders.

[7]*Fayol, H., Translated from French edition (Dunod) by Constance Storrs,* General and Industrial Management," *London: Pitman, 1949.*

(4) *Coordinate:* To harmonize all the activities of the firm—to schedule and sequence its affairs within an order of priority.
(5) *Control:* To report deviations from plan (monitoring) and to take steps to bring a situation back to plan (correcting); to verify whether everything is proceeding in accord with the plan and to rectify weaknesses.

Management operates within these elements through the personnel of the firm. In operating through personnel, there are, Fayol says, 14 general principles the leader will follow:

(1) *Division of work:* Foster specialization for productivity.
(2) *Authority:* Made up of personal authority and official authority. Authority is: earned vs. assigned, required vs. ascribed.
(3) *Discipline:* Obedience, energy, respect mutually accorded.
(4) *Unity of command:* Receive orders from one superior only.
(5) *Unity of direction:* One plan to reach one set of objectives.
(6) *Subordination of individual interests to the general interest:* No favoritism or private gain.
(7) *Remuneration:* Fair and satisfactory to firm and employee.
(8) *Centralization:* One central brain except in very large firms.
(9) *Line of authority:* Appropriate channels up, down, across.
(10) *Order:* A place for everyone and everyone in his place.
(11) *Equity:* Fairness and justice.
(12) *Initiative:* Maintain freedom to propose and inspire action.
(13) *Esprit de corps:* Harmony and union among workers and managers.
(14) *Span of control:* Supervision appropriate to the job.

THEORY: GEORGE MORRISEY

Management by Objectives[8]

The list of people preaching management by objectives would fill a volume in itself ranging from Drucker to Reddin and beyond. Morrisey's writings are, however, the most practical.

The manager follows a seven step process:

(1) *Defines roles and missions:* nature and scope of work.
(2) *Determines key results areas:* where to invest time, money and talents.
(3) *Identifies and specifies indicators:* determines measureable factors on which objectives may be set.

[8]*Morrisey, G. L.,* Management by Objectives and Results for Business and Industry, *Reading, MA: Addison-Wesley, 1977.*

(4) *Sets objectives:* Results to be achieved. Writing format is as follows:

To (action or accomplishment verb) a (single key result) by (target date) at a (cost-scarce resources, etc.).

(5) *Sets action plans:* Determine how to achieve a specific objective including steps such as the programming (time and steps), scheduling, etc.
(6) *Controls:* Measuring variances and correcting.
(7) *Communicating:* Working out objectives so they dovetail with those of others up and down the line.

USE: THE PROCESS When leadership is recognized as a series of processes, the manager will then enhance his/her self-development through readings in communications, planning, etc., and such overall systems as MBO. Besides building general skills, one or more such overall systems may be just what is needed for the operation. The system we briefly discussed, MBO, is a major aid to many engineering managers because engineering work is difficult to program on a day-to-day basis; thus, all personnel in a group must be continually reminded of the end product to which the organization or group is committed.

An MBO process is a key method for gaining subordinate participation in the management process. Whereas the seven steps make it look as if the manager sitting in splendid isolation does it all himself, the manager, in reality, dialogues each step with subordinates. They should be encouraged to frame their own responses at each step, offer their suggestions as to what should be framed for the group as a whole, and then they should dovetail their personal objectives with those of the group as a whole. There should be communication back and forth on what is framed and how.

THE EMPIRICAL SCHOOL: TRAITS OF LEADERSHIP

OVERVIEW EMPIRICAL SCHOOL The study of leadership as a subtle blend of characteristics leading to success has found favor with those interested in self-development. Each trait usually carries with it some self-evident paths or procedures for development. C. Northcote Parkinson[9] (formulator of Parkinson's Law) sums up the "traits school" by identifying six key traits.

[9] *Parkinson, C. N.,* The Law of Delay, *London: John Murray, 1970.*

ABILITY Defined as the aesthetic sense in fitting together materials and efforts so no one is idle and no one is overworked.

The difference between ability and skill:

skill: can do with ease what others find difficult.
ability: able to maintain the situation under proper control and the aesthetic sense.

One develops skill by practice. One enhances ability by thought, planning and the intelligent use of management systems.

KNOWLEDGE The information base a manager brings to a job and continuously enlarges while on the job. Components include:

(1) The general information base—the widespread background in the technical and managerial concepts germane to the field and industry.
(2) The specific information base—the background necessary for dealing with the firm's products and the management systems which would work in the context of that laboratory, research department, design department, etc.

The most serious difficulty the manager faces is rapid obsolescence the moment he is no longer engaged in hands-on technical work. Time is not sufficient to maintain state-of-the-art competence in technical areas as well as managerial ones.

The manager must have enough technical knowledge to audit, probe, question, and backstop a job.

1. Audit. To ask questions and understand the answers on the methods applied to yield technical developments. To be able to connect inappropriate methods with lack of accomplishment.

2. Probe. To be able to review a result, a decision, or an action by framing questions which uncover deficiencies or illuminate successes. The probe starts by the manager restating in his own words the meaning or intent of a subordinate's statement. From that, the probe can take either of two directions:

(a) A *confirmatory* probe. This is used to clarify facts, confirm facts, confirm attitudes, or enlarge on the basics. Example:

ENGINEER: "Although the circuit continued to overheat, we could not

find any component that was malfunctioning."
SUPERVISOR: "I see, then the currents in the circuitry were too high." (Clarifying a fact.)
ENGINEER: "We measured the currents and found them as calculated, but the transistors are running hot."
SUPERVISOR: "You are examining the heat sinks and the air circulation?" (Confirming the facts and enlarging on the basics.)

(b) A *leading* probe. This is used to suggest disagreement without actually contradicting the other person. When you want to persuade someone to reconsider, use a leading probe. Example:

SUPERIOR: "I think that the deficiencies in your operation can be traced to the engineers' lack of application to the job."
MANAGER: "You think this is true of each and every person in the group?"

3. Question. To be able to "turn on" subordinates' thinking when they are stumped; to use questions as prods, as challenges, as new directions for activities. There are two types of questions.

(a) Directive questions. These are narrow in scope and seek a specific piece of information; may often be answered with a "yes" or "no." A directive question focuses directly on the information item being sought. Example:

MANAGER: "Did you check specification X to see if our design conforms to the requirements?"

(b) Nondirective questions. These are wide latitude, broad in scope. The answer to such a question is limited only by the imagination of the respondent and opens the discussion to wider considerations. Example:

MANAGER: "Why do you think we are having difficulty meeting the specifications?"

How, what, when, where, and why are the foundations of most questions. How, when, and where considerations are usually nondirective. Be aware of why you, the manager, seek information—keep the "payoff" of that information in mind and formulate and reformulate a mixture of directive/nondirective questions and confirmatory/leading probes until you are satisfied with the facts/impressions you have obtained. Maintain receptivity to "broad" considerations and information even when the flow of information on the specifics appears adequate. The broad information may close loose ends or create an awareness of danger in the use of the specific information.

4. To backstop. To know technical capabilities of subordinates and understand the requirements of the jobs at hand; to assign appropriate primary resources and maintain appropriate reserve resources to get the job done.

Knowledge is achieved through readings, convention attendance, and college courses. **(Note:** Research shows that *casual* attendance at unstructured seminars makes minimal demands on concentration or effort and yields minimal results.) A manager should set up a comprehensive but *feasible* self-development program and stick to it. The half-life of an engineer is about eight years. The manager of engineers cannot have much more than that.

IMAGINATION Defined as the ability to perceive the needed end product and to communicate this perception to subordinates. Imagination is a key requirement for a functioning management-by-objectives program which in turn is both a stimulation and developer of imagination. It is also the ability to establish targets above the status quo.

DETERMINATION Instilling the belief that what can be done will be done:

(a) By describing the target to be achieved so that the group perceives it as worthwhile and attainable;
(b) So that there is a ''need'' by group members for accomplishment and a willingness to make personal commitments;
(c) While building the group's competence to achieve success in ever more challenging assignments.

This is a key area in which a manager must have a strong belief and feeling for the work of the organization. If this is lacking in you, a job change may help restore your commitment. A good management-by-objectives system is of help here.

RUTHLESSNESS This quality usually does not show up in managerial studies of organizations—the application of motivators, satisfiers and sundry techniques takes care of ''book'' problems. Parkinson feels that the strongest executives are those with the ability to rid the organization of those who are:

(1) *Disloyal.* Those who oppose for narrow or selfish reasons beyond the

boundaries of what the firm can live with (disruptions, public problems, etc.). The manager must be careful to differentiate between disloyal and loyal opposition—sometimes the devil's advocate plays a major role in preventing serious mistakes.

(2) *Careless.* Those who allow opportunities to slip by and pose inordinate opportunity costs for whatever accomplishments are shown.

(3) *Idle.* Those who have mistaken effort for results or are incapable of converting what efforts they do exert into accomplishments the organization needs.

(4) *Shelf sitters.* Those who live off their experience and do the minimum necessary to get by. They represent filled slots which might be better used by innovative or more accomplishment oriented people.

The ruthless manager is not afraid to instill stress in the organization in order to increase the rate of accomplishment. Stress will get people to work harder at what they know how to do; however, it interferes with learning. (Analyses of the effects of stress on innovation are not as definitive. The balance of sentiment leans toward the idea that some stress or dissatisfaction is a spur to creativity.)

An MBO system forces the establishment of documented targets; carried to the worker level, it lets management judge results on the basis of mutually agreed assignments to that person instead of unmeasurables. With such a system and audits of the methods being followed, a manager has the backup documentation for being ruthless.

Parkinson's harsh emphasis on ruthlessness is probably the product of the sloppy work we see all around us and the managerial acceptance of that work. The working manager must temper this harshness with a feeling of compassion for the people working for and around him or her; they are not machines and should be given room to explain, correct, or work around the difficulties that may be encountered. The value of the ruthless attitude may lie in an understated, but nevertheless very real determination on the part of the manager not to put up with substandard work or substandard performers. This attitude surfaces when people are given a chance, are shown every consideration, are encouraged to provide feedback, are listened to, but come up with excuses instead of results.

IMAGE Eugene Jennings[10], uses "visiposure" to describe this quality. A leader with image generates an atmosphere in which people want to use him/her as a role model. The leader places himself where the top brass can see him and appreciate the qualities of the group backing him. The leader goes out of his

[10] *Jennings, E. E.,* Routes to the Executive Suite, *New York: McGraw-Hill, 1971.*

way to expose the group's successes and to be visible when challenging assignments are "up for grabs." The leader is perceived to be playing a winning strategy—that he himself is on a winning track and will carry others along with him.

Self-image starts with elimination of flaws in such things as speaking ability, manners, tact, dress and proceeds to careful personal public relations (PR) built on performance and real powers.

SUMMARY

We have briefly exposed the reader to three views of leadership:

(1) as a behavioral pattern
(2) as a process
(3) as a blend of characteristics/traits.

We did this to:

(1) Encourage the reader to refer to the works of the authors mentioned and to study and learn more of the three views from a wide range of sources.
(2) To make the manager more aware. Symptoms of difficulty will often be masks for deep problems—inappropriate styles, lack of good processes, or weaknesses in key traits—which could be eliminated by early detection and action. A conceptual background allows the manager to "fit" pieces of a "puzzle" together quickly and more efficiently. A conceptual scheme in the manager's mind is a form of road map and provides direction instead of random action.

CHAPTER 4
Responsibility for the Bottom Line

Chapter 4
Responsibility For The Bottom Line

INTRODUCTION

There is a concept known as "Goal Congruency" which we can borrow from the social scientists to justify an examination of a manager's financial role in the firm. The firm expects that the summation of the goals of its component parts will coincide at all points with the goals of the firm as a whole. To assure this congruence, it is necessary for the manager to understand the impact of his goals (and actions) upon those of the firm especially in an area key to the firm's survival—finance. With this understanding, the manager can make better decisions which meet the intent as opposed to simply the letter of the firm's objectives. Even more important, understanding fosters initiatives: the manager need not wait for goals or "orders" to filter down from the top ranks of management. Financial indicators can be storm warnings or harbingers of opportunities which must be seized for immediate exploitation or they will be lost. The alert engineering manager can use them to guide himself in the deployment of available resources and to advance his own cause by doing more for the firm than is "required." He will be able to increase the firm's financial return on its investment in his operations.

This return, in the jargon of Wall Street, is referred to by its position on the accounting statement of income and earnings by which the firm determines how much money it cleared for a given period. In a firm's listing of earnings less expenses and other financial drains, the very last line, at the foot of the entire statement is an item marked net income (or loss) after taxes. The president of the firm is charged with the ultimate responsibility for attaining a net income which satisfies the firm's stockholders. His success, however, is predicated on the existence within each component of an appreciation of how that component impacts on the president's responsibility. This appreciation starts with the knowledge of the firm's financial machinery—the subject of this chapter.

PRICING AND COSTING

OVERVIEW

One of the major decisions a firm must make involves the price it will charge for its products. A budding entrepreneur will find that the first question financiers ask about the service or product he proposes to sell is, "What do you intend to charge?" Regardless of the technical wizardry involved, if the proposed price does not elicit a favorable "gut reaction," it will be tough sledding when he seeks to raise money. The established firm without a coherent, intelligently thought out pricing policy will soon find itself buffeted by decreasing profits or excessive competition; those firms engaged in bidding on procurement contracts are especially sensitive to pricing decisions.

Pricers must consider psychological, sociological, legal, economic, and cost factors. The psychological factor is most predominant in the consumer market; consumers have demonstrated a mind set about a "right price" for products ranging from ball point pens to consumer electronics. A lower price is associated with inferior quality and a higher price with a "rip off." Sociological factors are demonstrated by a firm's use of a high price model to raise the status of the entire line or the sale of a limited profit item because it "keeps" prestigious customers. Legal influences encompass antitrust laws (price fixing), anti price discrimination (a discount rate to one customer must be available to all, kickback-rebate prohibitions), provisions of contract law (the Uniform Commercial Code), and provisions of federal procurement agencies. The major influences we examine here are economic and usually cost related. Economic influences suggest the type of pricing decision a firm faces which depends, in turn, on the type of market in which the product or service is sold. Cost figures must be used in pricing to assure that revenues exceed costs. In some types of market, pricing techniques make little use of costs (women's cosmetics, for example) while in other types cost is the determining factor in setting the price (sealed bids, for example).

COSTING

Three common methods of costing are:

direct (or variable) costing
absorption (or conventional) costing
marginal (or incremental) costing

Which type of costing to use is an executive decision. Inventory valuation is most often done with absorption costing. Firms operating in extremely competitive markets may use direct costing for some lines and absorption for not so competitive lines.

DIRECT COSTING

Prime costs for the product are determined—direct material and direct labor—from the number of items produced, material usage records, and labor hour records. Sometimes these are obtained by careful questioning of supervisory and production workers who can provide a reasonable estimate. Next, the unit variable factory overhead cost is calculated. Variable overhead is the cost of such items as electricity, repairs, cleaning, and other items which vary with production. Fixed overhead includes such charges as depreciation, property taxes, and salaries. It is often quite difficult to decide if an overhead cost is fixed or variable; for example, salaried personnel might be let go with any downturn in work volume in some places (a variable cost), while other firms will retain them to do "advanced thinking" (treating them as a fixed cost).

The variable overhead must be allocated to the product on the basis of some activity measure accurately reflecting on the amount of overhead generated. Firms use direct labor cost, direct labor hours, units produced, or machine hours; the choice of which to use is aided by regression analyses.[11]

Variable overhead, direct labor cost, and raw material cost are then combined to arrive at a direct product cost figure. To this is then added variable selling and administrative costs; these costs are usually determined by preparing graphs of costs versus unit sales for various years and separating those which varied with sales from those which are fixed (stayed relatively

[11] *The figures are usually collected by the cost accounting function of the firm. Use of these figures in regression analysis consists of determining how the dependent variable "Overhead" varies in accordance with such independent variables as "direct labor hours," "raw material cost," etc. In mathematical terms, a dependent variable Y is said to be a function of the independent variable X. In statistical terms, however, the descriptive term "regression" is used in place of the words "a function of." Francis Galton, a London satistician of the 1880's, developed the idea of regression in his studies of inheritance. His friend, Karl Pearson, collected more than a thousand records of stature, cubit and span in family groups. His regression of son's height versus father's height showed that, as expected, tall fathers had tall sons (yet, interestingly enough, the average height of a group of tall sons was found to be less than that of their father's average).*

constant). The method of separating the costs will vary from firm to firm, but the combination of all the variables detailed provides the pricer with the information needed.

ABSORPTION COSTING

Absorption costing includes not only the direct labor, direct material, and variable overhead, but, also, an allocated portion of the firm's fixed overhead. As a general rule, the base used to allocate variable overhead is also used for the fixed overhead. Predicting the activity rate of the item used as an allocation base is a serious problem. If the activity is wide of the chosen measure, the per unit allocation may be badly under or overstated. Further, there are some accountants who feel fixed costs should not be allocated.[12]

MARGINAL COSTING

Marginal costing comes from the economic concept: it is the increase in the firm's total cost that results from increasing production by one unit. One finds the firm's total costs at varying levels of production and by taking appropriate differences derives the marginal or (in business terms) the incremental costs. The economist usually looks at marginal costs on a unit by unit basis; the accountant looks at incremental costs for broad changes in production. The accountant will usually start with all the direct cost data as the basis for arriving at incremental costs.[13]

PRICING

Pricing techniques fall under the following general classifications:

Straight Markup Pricing
Acceptance Pricing
Bid Pricing
Target Rate of Return Pricing
Marginal Revenue Pricing
Rental Pricing
Skimming Pricing

[12]*Accounting, as you may guess from this, is a dynamic consensus of the proper way to record revenues, expenses, and other costs. It is a consensus of officials from the largest accounting firms, academic accounting specialists, and officials from the taxing agencies of state and federal governments leavened by each firm's adoption of rules and rule interpretations which show its operations in the best light. Accounting "rules" are the antithesis of engineering "laws" which arise from the natural state of matter and energy.*

[13]*For reference, see* Pricing Techniques for the Financial Executive, *Wiley-Interscience, New York, 1974.*

STRAIGHT MARKUP PRICING

One determines the direct cost, sets the markup rate, converts the rate to dollars (by multiplying rate times direct cost) and adds this markup to the cost to arrive at the price. Absorption costing is generally used as the base. The more costs included in the base, the lower the markup rate. If selling and other administrative costs are not in the base, a higher rate will be used. Firms using straight markup usually choose not to allocate fixed costs. They set markup rates high enough so that a targeted volume provides coverage of the costs plus profits. This technique is the easiest and most common.

ACCEPTANCE PRICING

There are firms in a number of industries which individually can signal a price increase or decrease for everyone by announcing their own pricing decisions. Such an industry has "price leaders" and "price followers." The role of the price leader is to alert the other firms in the industry to the need for a price increase or decrease. Management in a price follower position needs accurate cost and activity projections. The follower examines his profitability before accepting a price set by the leader. The decision to accept or reject a price lead often contains elements of gamesmanship because the firm might be better off lagging behind as all others raise their prices or taking an intermediate price hike, etc. Price followerships exist when the major motivation of all the "players" is to avoid cutthroat competition and restraint-of-trade charges by the government.

Acceptance pricing becomes an even more complex decision when the product is such it can be bought "anywhere" and no firms control much of a market for it. Management can then: (a) accept industry prices, e.g., underprice to keep out (or even force out) potential competition; (b) reject the industry price and go it alone; or (c) shut down the plant and shift resources to a different product.

Firms must be extremely careful not to give the appearance of collusion in price setting or in any form of pricing below cost specifically to drive someone out of business. An injured firm or the government can strike back with antitrust charges and the penalties are very, very heavy if the defendant loses.

BID PRICING

Bid pricing is most important to companies developing a new product or dealing with government agencies. The buyer initiates the process by advertising a desire to buy a specific product by a specific date; the buyer may send out detailed specifications or a simple performance measure the product must meet.

A new product may be bid on either of two strategies. In the first one, a minimum bid is determined on the basis of all incremental costs—the costs include any items which will be needed only if the bid is won (including testing, setup, design, etc.) and no production past the bid contract period is scheduled. In the second strategy, potential users of the product in all industries are identified. They are surveyed re their needs and the extent of the market determined. From this, the firm estimates the activity base for overhead allocation, probable direct costs, variable costs, etc., and uses only the variable costs for the minimum bid to edge into the market. The desired profit is then added and the actual bid is set at some point above the minimum.

On older products, past experience will provide the cost data on which to set the minimum bid. Another way is to assemble a set of standard time and material measures to determine the minimum and to add some acceptable profit margin for the actual bid. How far above the minimum to bid is a question of real executive soul searching. There are risks inherent in taking on the contract, there will be investments to be financed because of the contract, and there are always alternative areas for expenditure of resources. Finally, there is one factor which is calculated to make life miserable for the manager preparing a bid. This is loosely termed "inaccuracy." Were the specifications interpreted properly or were they inaccurately read to call for something easy to do (when the opposite turns out to be true)? Were costs accurately calculated or were guesses used without much basis in fact? Was the correct rate of inflation taken into account? These questions often cause a split in management ranks. The legal department would like to write a contract that protects the firm against any of these inaccuracies and comes up with a contract that would serve admirably in buying gilt edged investment bonds—but wins no bid competitions. The sales department would like to be able to accept anything the customer has in mind—the objective is to get that contract into the "house" and then "we'll let engineering worry about the risks of noncompliance."

TARGET RATE OF RETURN PRICING

Under this form of pricing, one selects a desired rate of return on the investment (the assets of the business). The activity level for a reasonable period into the future is estimated (market research). The absorption costing method is used to determine unit cost at that level of activity. A price is computed to yield the targeted rate of return on investment at the expected level of activity.

The rate of return used must be sufficient to attract capital to the firm and to allow internal financing of expansion. The level reflects market share considerations, company image, and other factors management deems relevant. The key problems of this form of pricing lie in the realm of estimating future activity, the costs of capital, costs of expansion, the effects of competition, and the probability of attracting competition into the business where there is a high rate of return. (These are some of the considerations which require careful goal setting and policy decisions at the top from where these are visible.)

MARGINAL REVENUE PRICING Marginal revenue is the increase in total revenue resulting from the sale of one additional unit of output. In order to calculate what the effects of price changes would be on units sold, the firm will do test market surveys and regression analyses of past marketing data. In economic theory, the firm starting at a high price then keeps lowering price to increase unit sales and total revenue. The firm stops when marginal cost equals marginal revenue as this yields the largest profit for the firm. In practical terms, the firm usually wants to "protect" sales it is making at certain (higher) prices; where it feels it cannot isolate these from the rest of the market, the higher price may be the floor.

RENTAL PRICING Rental pricing is prevalent in industries which market products with a very high rate of obsolescence. Under these conditions, the rental is equivalent to a sale for a given period of time with a guaranteed buy-back by the seller. The steps for rental pricing include computing an absorption costing product cost; a markup is then calculated; the markup is added to cost and total revenue over rental life is calculated; then the total revenue is divided by the rental years to obtain the yearly rental. If funds are to be tied up for a long period of time, the rate of return must cover the financing charges as well as all the others. (This is the favored method for computers.)

SKIMMING PRICING A firm which develops something new, different, or significantly better than a product on the market can adopt what may be called a "skimming pricing" policy. Briefly, it consists of setting a very high price on the product so as to appeal only to the very rich or those who must absolutely have the product. When all those who can afford to buy have been reached, and this level of market exhausted (or skimmed), the price is lowered a notch to bring the "not so rich" into the market. The product may be given some cosmetic changes or some "deluxe feature" is deleted to keep former customers from getting upset. (In this connection, it is wise to note that here is where many product liability suits have their genesis. Something a cost-cutter thinks is deluxe or better than needed at the lower price turns out to be a necessity from a safety viewpoint. "Dollars above safety" is disastrous in the courtroom.) When the market at the new lower price has been exhausted (or, again, skimmed), the price is lowered once again. The major influence preventing every firm with a new product from practicing skimming is fear of encouraging competition to come into the market with a price that sharply undercuts it.

CONTRIBUTION TO PROFITS

In order to keep track of the source of overhead and to properly credit services or products with earned revenue, the firm may set up accounting entities known as cost and revenue centers. Whether or not the firm's accounting structure follows the reasoning we use here, it is a useful practice for the engineering manager to learn:

- The boundaries of his group as a cost center (i.e., is he part of a larger entity or a center unto himself).
- The boundaries of the revenue centers. If engineering is strictly a "fixed" overhead item, it may not be part of a revenue center, but attributed to the firm's cost as a whole. Some feel this structure cushions the department against revenue swings. The converse of this is that engineering may get undeserved cutbacks should general overhead rise beyond management's discretionary "limit."
- How the costs and revenues are allocated. It is usually not to difficult to learn about costs as they show up in budget charges. Revenues, however, may be calculated on a product line basis or involve a service encompassing the efforts of several major groups. It becomes a challenging exercise to take the revenue figures and to allocate them to a group so that the boundaries for costs and revenues match. (Get help from accounting.)
- Determine the group's contribution to profits—revenues less its budgeted costs. If the figure comes up negative, it may mean that the manager overlooked revenue from products or services which absorb engineering time, or his budget is being charged for time and other resources which should be part of a "basic research" overhead budget.

SOME BASIC ANALYSIS METHODS

COST-BENEFIT VS. COST-EFFECTIVENESS

A distinction may be drawn between the two types of analysis:

Cost-Benefit: Both items are in the same units. If costs are in dollars, benefits are in dollars; if cost are in man-hours, benefits are calculated in man-hours. Ratios usually have Benefits over Cost.

Cost-Effectiveness: The two sides of the analysis are in different units. Cost may be in dollars while effectiveness may be a composite rating reflecting papers published, inventories, problems solved, etc. Ratios usually have Effectiveness measure over Cost.

Many managers do not see the need to calculate their contribution to profits, or a ratio of benefits to cost, or to establish a measure of effectiveness until a top management order comes down abolishing their groups.

Do get your accounting people to help you determine a fair percentage of revenue or the portion of the firm's output you can allocate to your efforts. Do ascertain that you are being properly charged with overhead and other costs reflecting your group's activities (and that you are not subsidizing others not under your control). Do take a ratio of benefits to cost and do derive a figure for contribution to profits. Track this over time and obtain higher management advice and consent to what you are doing. Match your group's image (public relations) power to your performance power as appropriate. This means regular reports moving up, informal briefing to subordinates, and that an image of success is projected to all others.

Do derive a measure of effectiveness appropriate to your group. If you check the literature you will find widespread confusion on how to do this. This is mainly because each writer must keep a wide audience in mind when making suggestions. However, within your firm, you must determine:

(a) What measures management uses in evaluating the worth of engineering.
(b) How your superior judges the worth of the group.
(c) How similar groups are judged in other firms; getting this information is a key benefit of attending conventions, joining management societies, and partaking in person-to-person interaction at professional activities.
(d) What some of the measures in the literature are, including patents achieved, cost/schedule performance, design innovations, papers, awards, problems solved, and MBO (so you know when you have met an agreed upon objective).

Combine your measures, as appropriate, and use them as an effectiveness measure which you track on a weekly or monthly basis. Evaluate your performance and use it to build your image with higher management.

PROFIT AND LOSS STATEMENTS— BALANCE SHEETS

Numerous accounting books in the local library provide a detailed analysis of debits, credits, assets, liabilities, and their significance. What is attempted here is not as good as a trip to the library, but should be of assistance until then. Mention was made of a bottom line; in this section we will examine it more closely.

The income statement shows income, expenses, and net income (or loss) over a particular period of time—usually the fiscal period. The balance sheet lists all the assets, liabilities, and capital as of a specific date. The financial *progress* of a business is revealed by its income statement for that period whereas the financial *condition* on a specific date is revealed by the balance sheet. Anyone studying these documents can derive an excellent picture of the health of the firm; the usual technique begins with a "spread" of the statements. In a spread, each of the statements is placed side by side with those for five to ten previous years so that they may be viewed in a way which makes changes and trends stand out. Usually the influence of a given administration or management "reign" is limited to creating or changing a trend. Conceptually, the preparation of the income statement is quite simple. One lists all the revenue, subtracts all returns, damage, refunds, discounts, etc., to derive net revenue. Then expenses, the costs of the goods sold and other expenditures are subtracted and net income (or loss) before taxes is recorded. A subtraction is made for taxes paid and net income is shown.

The balance sheet shows assets (representing things owned by the firm) on the left of the sheet and liabilities (representing what is owed to others) on the right. A third account, placed beneath liabilities on the right side is called Net Worth. Net Worth reflects what belongs to the business or its owner after all the liabilities have been paid; net worth equals assets minus liabilities. Corporations will have such items as common stock, capital surplus, and retained earnings in the net worth accounts. As profits increase, the earnings the firm does not disburse as dividends go into the net worth accounts—balanced by stock holdings, cash, bonds, etc., on the asset side.

The firm's results re earnings per share (net after tax divided by number of shares outstanding), and percentage of market (the firm's sales as a percentage of total industry sales per year) plotted as trends will generally reveal how well things are going for it compared to other firms. Government agencies are demanding that firms break out their earnings by major lines of products when issuing year-end reports; thus you may be able to get a significant reading about the results and trends of major competitors. Should your trends go down while theirs go up, you may want to launch a careful study of their contract methods, sales strategies, and the characteristics of their products.

The manager who wishes to develop a profit oriented outlook might find it quite useful to analyze his firm's income and balance sheets. By examining the trends, he might pick up substantial clues to the profitable investment of engineering efforts well before "orders come down" from the top. Tracking the figures and taking appropriate action to counter negative trends can provide a manager with proof that his goals are congruent with those of upper management.

The manager should also determine the criteria the firm uses to judge new ventures, new product development, and similar investment opportunities; such knowledge is of immense benefit when the manager sits down to write an investment proposal or justification for new capital funds.

THE BASIC ANALYSIS TOOLS

The basic tools for analysis are generally available to anyone who digs for them. They include:

(1) Industry performance figures put out in various publications of Dun & Bradstreet including information on sales, costs, profits, etc.; *Barrons, Business Week* and industry trade associations are also good sources.
(2) Profit and Loss (Income) Statements for the last five to ten years. The same statement might also be available in a percent of sales format; if not, this may be calculated by the manager using net sales as the 100% base.
(3) Balance sheets for the 5–10 year period.

ANALYSIS P & L

This analysis is designed to help the engineering manager determine where the opportunities for his/her initiative lay.

At the top of the list of concerns is sales. Using the industry figures and the company figures, construct a table of percentages using a convenient year as a base of 100. Convert the industry figure on its base and repeat for the company figures on its base in the same way. Then examine the year-to-year percentage changes for industry and the year-to-year percentage changes for your firm.

If the industry grew at a better rate than your firm, your firm lost part of its share of the market. Check to see if:

(a) Competitors' product features are better.
(b) Competitors' products underprice yours.
(c) Reliability and maintenance cost of competitors' products are less.
(d) Your firm's quality control is not as good as that of your competitors.

Talk to your firm's marketing, advertising, sales, and maintenance personnel; get out and talk to customers and find out how the product is doing. Take the initiative! Check the income sheet sales trend. If down, see if, in addition to the above, the following exist:

(e) Product safety deficient (including its instructions, labeling, warnings, design, and construction).
(f) Production difficulties are holding up deliveries.

Convert the profit and loss statement dollar figures to a percentage of net sales figures. Check the trend in percentage of sales returns and allowances. An upward trend may be due to:

(g) Poor quality control.
(h) Excessive sales puffery resulting in unfulfilled expectations.
(i) Product linc prolifcration and customer confusion.

Next, check the cost of goods sold as a percentage of sales and note the trend.

A rise in unit labor costs should suggest these avenues of investigation for you. See whether:

(j) A change in manufacturing processes, product simplification, or tool redesign will allow reduction of labor content or the use of lower skilled (cheaper) labor grades.
(k) New machines on the market can save enough labor content to yield a desired return on investment. Maintain an open door to salesmen who will bring you new ideas. Temper your enthusiasm for any new device with the realization that the firm's investment schedule might have higher priorities (or investment funds may simply not be available regardless of their expected "earnings").
(l) Production can be highly automated and/or switched to off-shore manufacturing facilities.
(m) Additional manpower training is needed in assembly, etc.

If materials costs have gone up, check to see if:

(n) Materials substitions are feasible. Before approving any changes keep in mind that a material which is a substitute, and fails because it has not had the vigorous testing of the original, can be the basis of a damaging product liability suit. Substitute if you can, but make sure you cut no corner on testing and evaluation of every new material or component.
(o) Production changes can make better use of what is now scrap.
(p) Value engineering can take the "fat" out of the product.

If overhead costs are up, check:

(q) Machine downtime and service methods.

(r) Proliferation of multiple specifications for products which are almost identical except for one or two features; combine features where possible.
(s) Unneeded space which is being unnecessarily charged to the product.
(t) Proliferation of inspectors, expediters, and other staff because of incoming materials deficiencies, production problems, and deficiencies in the product design itself—making it difficult to turn out acceptable goods.
(u) Packaging of material—changes in product may reduce shipping costs, e.g., removing a handle (for customer assembly) may allow a sizeable package reduction.

ANALYSIS BALANCE SHEET

The balance sheet reveals three major figures of interest to the engineering manager.

Raw Material Inventory

A rising trend may indicate:
(a) Company buying in anticipation of price increases or shortages.
(b) It is time to seek materials substitutes (reinforcing ideas from P&L).
(c) An obsolete part, eliminated by a design change, is still being ordered because of a lag in changing parts lists for purchasing.
(d) Inflation may have removed or reversed a "favorable" cost spread between two competing suppliers and your firm's buying patterns have not been altered appropriately.

Work in Progress Inventory

A rising trend may indicate:
(e) Company reevaluations of the dollar value.
(f) Production bottlenecks caused by poor product design, poor plant layout, machine problems or too long a process time.

Finished Goods Inventory:

A rising trend may indicate:
(g) Company dollar reevaluations of the inventory.
(h) Poor shipping practices due to packaging, product proliferation, product damage/spoilage.

EXAMPLE

Income (P&L) Statements, Balance Sheets, and Industry figures are obtained for Jones Widgets for three years; see Table 4.1. (Normally, analysis of ten years is better.)

TABLE 4.1

• *Profit & Loss for Jones Widget (in thousands of $)*

	1970	1973	1976
Sales	1000	2000	3000
Sales Returns	50	100	200
Net Sales	950	1900	2800
Cost of Goods Sold (Mfg. Cost)			
Materials	150	300	600
Labor	100	250	400
Mfg. Overhead	50	100	200
C.G.S.	300	650	1200
Gross Profit on Sales	650	1250	1600
Operating Overhead			
Selling & Adv.	100	300	500
Admin.	50	150	250
Op. Overhead	150	450	750
Net Income before tax	500	800	850
Taxes (50%)	250	400	425
Net Income after tax	250	400	425

• *Balance Sheets (Partial) for Jones Widget (in thousands of $)*

	1970	*1973*	*1976*
Assets			
Cash	500	800	1000
Accounts Receivable	200	400	500
Inventories			
Raw Materials	1000	3000	4000
Work in Process	600	800	1500
Finished Goods	1200	1800	2500

• *Industry Figures for Mfg & Sales of Widget (in hundreds of thousands of $)*

	1970	*1973*	*1976*
Sales	*300*	*900*	*1800*
Mfg. Costs	*75*	*200*	*375*
Advertising & Admin. Overhead	*25*	*50*	*225*
Profits after Taxes	*100*	*325*	*600*

Analysis

Examine the increase in industry sales versus your firm's sales on an index basis (using 1970 figures as 100%, divide the dollar numbers).

Year	Industry index	% Change	Jones firm index	% Change
1970	100	—	100	—
1973	300	+ 200	200	+ 100
1976	600	+ 100	300	+ 50

Widget industry growth has outdistanced the Jones firm by a considerable percentage. Jones firm has lost share of the market as below:

Year	Industry Sales × 10^5	Jones Net Sales × 10^5	Jones % of Market	% Jones Change in Share
1970	300	9.5	3.2	—
1973	900	19.0	2.1	−34%
1976	1800	28.0	1.6	−23%

Jones firm should check items (a), (b), (c), (d) (under P&L Anal.)

One can also check manufacturing costs as a percentage of (net) sales.

Year	Industry Mfg/Sales	Jones Mfg/Sales	% Change in industry	% Change Jones
1970	25%	31.5%	—	—
1973	22%	34.2%	−12 %	+ 8%
1976	21%	42.8%	− 4.5%	+25%

While the industry cut its manufacturing cost per sales dollar, Jones has had enormous increases. The dollar sales trend is up, but not as much as market growth. Check items (e) and (f).

Convert the P&L to a percent of sales sheet (rounded to nearest whole); see Table 4.2.

TABLE 4.2

Jones Widget P&L (in % of Net Sales)

	1970	1973	1976
Sales	105	105	107
Sales Returns	5	5	7
Net Sales (used as base)	100	100	100
Cost of Goods Sold			
Materials	16	16	21
Labor	10	13	14
Mfg. Overhead	5	5	7
C.G.S.	31	34	42
Gross Profit on Sales	68	65	57
Operating Overhead			
Selling & Adv.	10	15	17
Administration	5	7	8
Op Overhead*	16	24	27
Net Income before Tax	52	42	30
Tax	26	21	15
Net Income after Tax	26	21	15

*Does not add because of rounding calculated from dollar figures.

Sales returns jumped in 1976. Check (g), (h), (i).

Cost of goods sold has had a sharp increase. Check (j), (k), (l), (m) for possible labor savings.

Materials costs have gone up. Check (n), (o), (p).

Manufacturing overhead is up. Check (q), (r), (s), (t), (u).

You may also wish to find out just why the operating overhead has gone up. Are the salesmen having problems "pushing" the product? Talk to them and see if you can help.

The balance sheet trends can be discerned on the basis of the dollar figures, but if the figures are normalized against each year's net sales, the results are as given in Table 4.3. The work in process inventory should be examined for items (c) and (d) on the balance sheet analysis.

TABLE 4.3

Year	Raw Materials	Raw Materials to Net Sales % Change	Work in Process	Work in Process to Net Sales % Change	Finished Goods	Finished Goods to Net Sales % Change
1970	1.05	—	.6	—	1.26	−25%
1973	1.58	50%	.4	−33%	.94	−25%
1976	1.43	− 9%	.5	25%	.89	− 5%

Example: 1970 R.M./Sales = 1000/950 = 1.05

1973 R.M./Sales = 3000/1900 = 1.58

% Change 1973 - 1970 = (1.58 - 1.05)/1.05 = .53/1.05 = 50.4 = 50%

BUDGETS

OVERVIEW Budgeting is usually thought of as a planning process—to relate the planned expenditure of funds to the accomplishment of planned objectives. However, the management of ongoing activities and the control of spending are two functions which have usually been given priority over the planning function. Every budget system comprises planning, management, and control processes.

Planning, in the budget context, is deciding on objectives, the evaluation of alternative courses of action, and the authorization of funds for select programs or activities.

Management, again in the budget context, is the process by which resources are obtained and used effectively and efficiently in the accomplishment of the objectives.

Control is the process, in the budget context, of assuring that specific tasks are carried out effectively and efficiently. Control binds operating officials to the policies and plans set by their superiors. Control is predominant during the execution stage of the budget process, but most frequently, the very form of budget estimation, budget appropriation, and reporting procedures set up during planning are determined mainly by control considerations.

YOUR ORGANIZATION'S BUDGET SYSTEM

To determine the budget orientation in your firm, examine the system being used.

(1) A planning orientation focuses on broad programs. It forces inclusion, on the budget forms, of statements on the goals of the program, priority analyses, relationships among programs, criteria for program allocation, and other decision oriented considerations.

(2) A management orientation focuses on costs for alternative staffing, costs for alternative engineering approaches to specific problems, efficiency and effectiveness measures to be applied.

(3) Control will accent expenditure ceilings, rates of expenditure, limits on expenditures without higher approval, and systems for the actual allocation of funds to the smallest unit of the department (one man, one instrument, etc.).

Every budget system will contain all three elements; the differences from system to system will not be the absence of any one element but rather the degree one element is emphasized. The usefulness of the knowledge becomes evident when a manager avoids trying to do too much with the system supplied by the firm. If you need a planning instrument and the firm supplies a control dominant system, you will have to recast what you do for the organization before it will work for you individually. Further, different personal skills are required in dealing with the complexities involved. A planning budget requires a good economics background and cost-benefit orientation. A management oriented budget requires a complete understanding of the administrative operations of the firm and the costs of resources on a program-wide basis. The control budget calls for accounting skills.

In a planning budget, the information focuses on the purposes of the program, and the key budget stage is in the pre-preparational activity where policy decisions are made.

In the management budget, the focus is on activities; the key stage is the preparation itself where efficiency is watched.

In the control budget, objects are tracked (wages, item expenditures), the key stage is execution, and the manager is really a fiduciary arm of the firm.

Setting up and using such a multifaceted tool as the budget calls for great skill in dealing with people. Remember that it takes different skills (language and sensitivities) to deal with accountants as opposed to economists as opposed to administrators as opposed to, finally, engineers. But doing so, successfully, is a prime requisite for becoming a well-regarded engineering manager (and more).

FIRST-LEVEL MANAGEMENT BUDGETS

At the first level of management, budgets are invariably control dominant; the larger the firm the more likely are the planning and management aspects to have been deeply subordinated.

The budget as an end product summarizes:

(1) Decisions for the future. When the budget detail is by project, or geared to a specific output item, it is a programmed budget. Most budgets are the traditional line items budgets for personnel, materials, and allocated overhead.
(2) The information relevant to the planned financial expenditures. The budget should allow all charges against the department with explanations and means for tracking them as time passes.
(3) Analytic concepts by which overhead is allocated.
(4) The administrative structure through which the budget will be administered, controlled and appraised.
(5) Justification for all planned expenditures.

Budgets will show the following types of information:

(1) Highlighted: The total dollars budgeted.
(2) Total dollars budgeted broken down by major classification of expenditure such as:
    Subcontracts
    Operations expenses
    Capital expenditures
    Rentals/leases
(3) Budgeted dollars with charges allocated by the firm's cost mechanisms.
    Gross Cost—Total cost incurred by your activity including charges for services rendered by other departments within the organization.
    Charge Outs—Costs your department charges to others for services you have provided.
    Net Cost—Gross costs less charge outs.
(4) Detail Costing
    Wages
    Salaries
    Fringe Benefits
    Data Processing
    Overhead dollars
    Supplies
    Travel
    Publication expenses.
(5) In a "rolling" budget, calculations of expenditure levels may be done for three years in advance with each current year moving into the year one position. Such a system allows careful monitoring of year to year expenditures for given projects—it adds a "rear view mirror."

Most firms require two separate budgets to be prepared: a Capital or Facilities Budget and an Expense or Operations Budget.

CAPITAL BUDGETS

The Capital Budget shows equipment dollar needs over the next three to five years (or more). A capital budget is prepared by listing all required instruments or specialty items in the order of priority needed. A few telephone calls will usually elicit rough costs for each item; a dollar sum for the next fiscal year is totaled for all equipment to be purchased in that fiscal period. If this sum exceeds that allowed by management for capital expenditures, for the year, the priority listing will have to be reshuffled. Each item must have full justification including, where appropriate, a demonstration of dollars saved by use of the equipment, and time to full payback (the savings are enough to repay the firm's investment plus all financing costs). Management guidelines are followed on how big an expenditure must be for an item to be capitalized (or how long a period of service it will render to be amortized).

EXPENSE BUDGETS

Expense budgets are usually prepared on an annual basis to coincide with the firm's fiscal year. The simplest form of budgeting starts with a list of items to be purchased, positions to be funded, total yearly expenses to be allocated to the department (floor space, heat, light), the total expenses expected for telephone, duplicating, and other administrative expenses, and expected expenses for internal central services such as data processing, printing, etc.

Theoretically, the sums allocated could be established as expenditure ceilings and the manager called to account for overruns only at the end of the year. The deficiencies of such a scheme are obvious.

The sums are allocated in either of two ways: By accrual or by actual expenditure. Under the first scheme, the total is allocated in even monthly installments over the relevant period within the year. Even if the bill, for example, of a service is charged quarterly, the expenditure for that service will be allocated monthly. Under actual expenditures, the dollar amounts are shown in the months the bills will actually come due and have to be paid even if on an internal accounting basis.

The usual system is for the firm to set up a series of computer account numbers and to charge them monthly for the services used. When the budget is prepared with all costs allocated on a month to month basis, it becomes easier to check the differences (or variances) of the planned expenditure against the actual expenditure. If a particular expenditure has a skewed funding profile, this can be entered so that artificial variances do not mask real ones. Variance analysis is the process of tracking down the reasons for going over or under budget. It usually is done in two parts. In the first, an analysis is made of all the price increases suffered per unit of labor, overhead, materials, and other components entering into the gross figures. The second analysis extracts the nature of the overruns in the numbers of hours used, units of materials consumed, etc.

VARIANCE EXAMPLE

Planned Project: 500 hours of engineering at $10/hr. Expected Cost: $5000.

Actual Result: Total Cost $9000.

Variance Analysis: Engineering rate went to $15/hr. Actually used 600 hours.

Cost overrun due to inflation: $\$5 \times 600 = \$3000$

Cost overrun due to hour overrun: $\$10 \times 100 = \$1000$

Total overrun $= \$3000 + \$1000 = \$4000$.

The manager is now faced with the task of tracking down the reasons behind the overrun. The increased engineering rate may have been company mandated due to general inflation in benefits, taxes, and other overhead items. Check to see that they were appropriately allocated. The overrun due to hours is more serious as this may be directly attributable to shortcomings of the manager himself. Examine the original estimate and the premises on which it was based. Did the job get all the physical and intellectual resources which had been promised? Were time and talent siphoned off in individually insignificant (but collectively disastrous), overlooked, ''quickie'' projects or ''emergencies''? Did the talent working on the job give its ''all''? The correct explanation of this overrun should lead to obvious remedies.

CAPITAL INVESTMENTS

Before a firm will invest its funds, it will ask the requisitioning manager to justify the need; how will the firm be better off investing its funds in a piece of equipment than in the alternatives open to it?

Most justification methods originate in simple discount cash flow analyses wherein the rate of return on the investment (ROI) is then compared to ROI for the alternative.

Using simple interest, one dollar placed in a bank at 5% per year interest will be worth $1.05 one year from today.

Consider a sum P at an interest i.

Sum at end of year $= P + Pi = P(1+i)$

At end of year 2, $S_2 = (P + Pi)i + (P + Pi)$
$= Pi + Pi^2 + P + Pi$
$= P(1 + i^2 + 2i)$
$= P(1 + i)^2$

At end of year n, $S_n = P(1 + i)^n$

Now, if we are promised a sum S to be given to us at the end of a period of n years instead of today, how much would we be willing to discount that sum to get it into our hands today—where we could invest it at our own discretion at interest i?

$$P = S_n/(1 + i)^n = [1/(1 + i)^n]S_n = DS_n$$

The factor D is the discount cash or present value factor and is available to us in various banking tables for complex interest periods, compounded, etc.

A SOURCE FOR INTEREST TABLES

Request the free booklet "Compound Interest Tables" (single copies only) from:

Corporate Communications
4th Floor
Union Carbide Corporation
270 Park Avenue
New York, New York 10017

Some calculators (e.g., TI's Business Analyst) can also perform the necessary calculations.

CAPITAL INVESTMENTS CONTINUED

Using the tables, we can find the interest rate, or in business investment terms, the return on investment.

In calculating return on investment we must always be sure to have a flow of cash or yearly earnings which:

(a) Cover the depreciation of the item.
(b) Return funds to firm at an interest on that investment equal to or better than can be obtained with alternative investments over the life of the item.

INVESTMENT EXAMPLE

A capital request for $20,000 to purchase and install a new small computer: Labor and other savings with this computer close to engineering (instead of using the central machine) estimated at $7,000 per year. Corporate tax rate of 50%, depreciation over 10 years. Minimum required rate of return is 20% after tax.

	Gross Savings	$ 7000
	Book Depreciation	2000
	Taxable Saving	$ 5000
	Tax (at 50%)	$ 2500
	Cash Flow	
	Gross Savings	$ 7000
	Less Tax	2500
Cash saving after tax		$ 4500

The Net Cash Receipts are discounted to zero date using the present value tables.

To "enter" the present value tables at an appropriate interest rate, calculate an initial trial rate:

$$\text{Initial trial rate} = \frac{\text{Average annual net cash in first 5 years}}{\text{gross outlay}}$$

So our trial rate = 4500/20000 = .225 = 23%

Try 25%

Set up data as shown in Table 4.4.

TABLE 4.4

Time period	Net cash receipts	Factor at 25%/yr	Discounted (present value of receipts
0	−20000	1.00	−20000
1	4500	.800	3600
2	4500	.640	2880
3	4500	.512	2304
4	4500	.410	1845
5	4500	.328	1476
6	4500	.262	1179
7	4500	.210	945
8	4500	.168	756
9	4500	.134	603
10	4500	.107	481
			− 3931

What we have done is to discount the value of the savings over the years by the factors of the trial interest rate, we then compared the total of the present values of the savings to the present value of the cash outlay at time 0. If we had the correct rate, the total would have been zero instead of $3931 negative. The negative indicates we discounted too much. So we can jump down to 20% and see what the table tooks like (see Table 4.5).

TABLE 4.5

Time period	Net cash receipts	Factor at 20%/yr	Discounted value
0	−20000	1.000	−20000
1	4500	.833	3748
2	4500	.694	3123
3	4500	.579	2605
4	4500	.482	2169
5	4500	.402	1809
6	4500	.335	1507
7	4500	.279	1255
8	4500	.233	1048
9	4500	.194	873
10	4500	.162	729
			− 1134

We could reject the investment out of hand because we would not be able to show a positive value summation at 20%. Let's try 15% and see what its summation shows.

Time	Net cash per year	At 15% Σ Factors 0–10	Discounted
0	–20000	1.000	–20000
0–10	4500	5.019	22585
			2585

We can interpolate:

$$\begin{aligned}\text{Actual ROI} &= 15\% + [2585/(2585 + 1134)](5\%)\\ &= 15\% + [2585/3719](5\%)\\ &= 15\% + 3.5 = 18.5\%\end{aligned}$$

MAKE OR BUY Firms will also ask for an analysis of whether it is cheaper to purchase or to build an item "in-house." Some points to analyze include:

(1) Check in-company estimating procedure.
(2) Check specifications given to vendors vs. those in-house.
(3) Has vendor given a complete quote?
(4) Vendor experience on this item.
(5) Can commitments made in-house or by vendor be considered realistic—what conditions are attached?
(6) Are there supporting tools, fixtures, etc., needed?
(7) Specialized training needed?
(8) Does vendor have better support capabilities in this specific area?
(9) Have all legal ramifications of the vendor-company deal been cleared?
(10) Is this project a profitable use of executive and other time if done in-house?
(11) What is the impact of poor or late performance on the prime objective?
(12) How much supervision will be required for a vendor vs. in-house?
(13) Is it better to buy or lease the necessary support facilities?

CLUES TO CORPORATE STRATEGY—PROFITS

STRATEGY OVERVIEW The success of any operation is predicated on a carefully planned allocation of time, funds and executive attention. The plan must contain rational guidelines spelling out the priorities, the relative emphases, and the time sequencing of the firm's initiatives in the marketplace, its internal operations, and its sources of supply. The decisions expressed in this plan are termed strategy. A manager can frequently discern the outlines of his firm's strategy by examining its view of profits (and how it allocates its resources to earn them).

Profits may be viewed as a reward for bearing risks and uncertainties. (Risk is differentiated from uncertainty by the fact that probabilities can be determined for the former, but not for the latter.) Under the risk and uncertainty doctrine, return on investment is the key to all decisions. The minimum investment needed to attain an acceptable cash flow must also yield profits at a rate of return that meets or exceeds very tight company guidelines. Firms whose sole criterion is return on investment, minimize expenditures on anything that is not immediately revenue producing (generating a cash flow). Customer service, quality control, technical backup do not yield cash flows and are "good" targets for minimum investment policies.

Profits may also be viewed as a consequence of imperfections in the marketplace. Such firms constantly strive to exploit consumer ignorance, fears, emotional hangups, or dislocations due to supply or transport. Such a management frequently strives for market control and tries to attain a take-it or leave-it position vis-a-vis its customers. This doctrine has led to the dynamics of market segmentation such as in autos where a firm appeals with specific nameplates to specific levels of income—although only the skin might be different from one level to another. Only rarely, in such a situation, are the users' real needs and expectations taken into account by the firm's marketing decisionmakers; their challenge is to sell features yielding the greatest profits rather than of educating the market to lasting worth.

The third broad class of profit seekers sees profitability as a function of innovation. This type of firm responds to market challenges as outlets for something better (in the truest sense of the word). It will respond to initiatives from its research and development to involve the user in something that will serve the user well at lower cost, greater speed, etc., than previously possible. The challenge of this firm's marketers is to educate the market to the true worth of the innovations being brought into being by the firm.

USE OF STRATEGY INFORMATION

It should not be difficult for the manager to see where his or her firm sits. As extreme examples, compare the degree of innovation between the auto industry and the computer industry. The first capitalizes on market imperfections; the second has a sophisticated buyer/user and must be oriented to the third profit philosophy. The impact on the degree of innovative engineering work that the industries are carrying on becomes obvious.

The knowledge of the firm's strategy enables the manager to formulate his goals so they become congruent with those of the firm. An engineering manager in a marketplace–imperfection oriented firm will probably be appreciated more for a flashy change than for something that makes the product last longer or work better than those of the competition. Conversely, in other industries, the manager who is content with seeing his group put out minor improvements or even minor cost reductions will one day wake up to find his whole "improved" product line is made obsolete by a different firm's innovative new-twist.

To prevent surprises and to lead your group in accordance with your firm's strategy, examine its profit view, and make yourself familiar with its products, distribution channels, pricing, and other external considerations; do the same for its asset distribution—which product lines get what share of production machinery, space, executive attention, and general efforts.

COMPETITIVE ADVANTAGES

A firm will have significant competitive advantages over other firms because of its engineering capabilities expressed as:

(a) Excellence of product performance or technical creativity or solutions arrived at for customers.
(b) Skill and low cost of operation of specific technical facilities.
(c) Industry leadership in innovation.
(d) Capability of amassing in-depth engineering brainpower which has kept up to date in its field.

The company may also have unique competitive advantages due to:

(a) Highly efficient production facilities.
(b) Access to scarce raw materials.
(c) A well-defined label in a strong distribution chain.
(d) A patent or specialized know-how.
(e) A strong financial position.
(f) A strong position in allied product markets.

It is also useful to track engineering activities to see if they are adequately servicing the firm's customers or helping the firm maintain its advantages by:

Customer Category—If the firm sells exclusively to the military can it make money by adding domestic markets? Are customers concentrated by small or medium or large firms? Can engineering help penetrate an unsold market?

Product Category—How is the product used? Can its application be stretched to do more? Can it substitute for other items?

Channel of Distribution—If a product is sold to home contractors, can it be made suitable, by some innovative engineering, for the do-it-yourself market? Or vice versa? If the product requires detailed follow-up, can engineering build in self-corrections?

Sales by Price/Quality—Can engineering help the marketing people up-grade the image to reach a high end of the market or delete certain features to penetrate the low end?

Finally, it is useful to be aware of the firm's policies in regards to dealing with:

(1) Consumerism and product safety
(2) The environment
(3) Ethics
(4) The political environment.

PERFORMANCE EVALUATION

OVERVIEW

The most difficult part of the manager's job is to evaluate fairly and competently the true accomplishments of an engineering group. One has to avoid confusion among:

Background—The sum of experiences, education and training a person has.

Capability—The experiences, education and training which a person can bring to bear on a problem—the applicable information base or mental superstructure—and the inherent motivation to apply it.

Skill—The ability to do with ease what others find difficult—sometimes labeled proficiency (how well one works).

Performance—The tangible, beneficial results of effort applied in accordance with the company's modus operandi.

Personality—The outward manifestations of a person's feelings, desires, methods of reaching internal goals.

Effectiveness—The degree to which a desired result has been achieved.

Efficiency—The value of output achieved for the value of input.

Effort—How hard a person works—intensity of application.

Skill, capability, and background may help win contracts, but performance, effectiveness, and efficiency determine the firm's success.

EARNED VALUE

The Department of Defense is pushing the concept of earned value to help managers keep projects on schedule.

Earned value is a derivative of Frederick W. Taylor's *Scientific Management.* Taylor espoused "task management" in which every piece of work was planned ahead and payment made on the value of the work achieved.

On the job, the manager and subordinate plan a "fair" month's work for a "fair" month's budget. The difficulty starts with defining the visible result the manager will see in order that all concerned really can agree work has been done. The fact that an effort has been applied is not a substitute for a check of tangible evidence that work is being accomplished. It is up to the manager to look at unsuccessful results, paper statements of the problem, difficulties logged, and other outward manifestations of the application of effort, and differentiate these from performance. The manager should talk to the individuals working on the project and note just which aspects of it give them satisfaction. Success oriented people focus on the results they seek or have attained; others take great pride in the efforts they have expended. Because it is easier to talk about efforts than unachieved results, too many groups become task oriented as opposed to results oriented. The performance of procedures and the hustle and bustle of activity are confused with satisfactory accomplishment. It is up to the manager to point out what is happening and get people to finish a project instead of playing with the test equipment—to buckle down on doing the project instead of the thousand and one diversions which can soak up time. Concentrated mental activity directed at the solution of a problem calls for intense effort which requires, in turn, a strong motivating force. If the work itself does not provide the force, the manager must do so and keep up an appropriate directed pressure to get things done.

After the performance is examined, a claimed earned value (dollars, hours, etc.) for that accomplishment is matched to the actual costs and the planned costs. Differences between actual and planned outcome values are variances.

The analysis of unfavorable variances must be written and each current task must be examined for estimated total cost at completion.[14] Overruns are immediately picked up and a poor variance trend may call for restructuring of original plans. The variance analyses should contain plans for future resolu-

[14] *This requirement has stimulated some of the most imaginative prose of this century.*

tion of any problems and the prevention of recurrence. The steps are then:

(1) Define work to be accomplished. A good MBO plan is extremely valuable here.
(2) Set up the accomplishment plan to include dates, budgeted dollars, and the potential value of accomplishment.
(3) Define the value measure for accomplishments.
(4) Perform variance analysis at appropriate periods in writing.
(5) Trace problems to sources and set up written plans for corrective actions.

PRODUCTIVITY MEASURES OVERVIEW

A new manager making a survey of the literature for productivity measures will soon suffer a mounting disillusionment: there are hundreds of books written about motivating people to be more "productive," articles discussing the various facets of productivity, and advice on being more efficient. But when it comes down to the individual manager attempting to evaluate the mental output of a group, the authors slickly write around the question. There really is no set answer to evaluating the output of a design group versus a blue sky research group versus a proposal group, etc. When it comes to judging their activities, a "gut" feeling of an experienced executive may be more accurate than all kinds of academic measures. But we are not completely without guidelines in forming these gut evaluations.

There are three basic concepts:

(1) We draw from decision theory the principle of using many judgment factors: The probability of being wrong is reduced, the criticality of the individual judgments decreases, and the time to decide increases as the number of factors evaluated for that decision is increased. A decision on the quality of a person's performance based on an overall view has far less probability of being right than a decision based on a summation of a series of separate judgments on a half dozen (or more) observed specific items.

(2) Every measure must be accompanied by a series of check and balance measures. This is to prevent anyone from working to satisfy a single performance measure at the expense of all other requirements of the firm.

(3) There are activities and results not amenable to a measure. There is nothing wrong with a manager expressing his or her feelings about these intangibles. The manager is responsible for a smooth operation and its results. If something generates negative feelings, or makes it difficult

for the manager to function with an individual, the point should be articulated and evaluations expressed even if it is not by a recognized yardstick.

MANAGEMENT BY OBJECTIVES

A good MBO system helps set the tone for a good evaluation system. Review this manual's section on MBO.

MEASURES FOR INDIVIDUALS

Suggested performance measures for individuals:

(a) Number of proposals written
(b) Complexity of proposals (simple, difficult, etc.)
(c) Number of letters written
(d) Complexity of matters handled
(e) Number of papers
(f) Number of internal reports
(g) Complexity of internal reports
(h) Degree of innovation expressed, demonstrated
(i) Professional awards
(j) Number of completed projects
(k) Number of reports
(l) Degree to which keeps at state of the art
(m) Colleague evaluation (judged by degree of interaction)
(n) Degree to which meets objectives
(o) Cost and schedule overruns
(p) Ingenuity in working around procedural problems
(q) The smoothness with which things proceed because of good planning.

Keep a log of your evaluations of the reports, drawings, etc., which cross your desk for approval. In this way you will have a year long record on which to form your opinions. Most engineering managers become conscious of their evaluation duties a week before the appraisal interview; such a manager formulates his judgments off the top of the head knowingly or unknowingly unduly influenced by a single recent fiasco or outstanding feat—the "what have you done for me lately?" syndrome. The employee will, naturally, bring forth a dozen examples to bolster or defend his side of the case and soon all pretense of evaluation is lost in a destructive give and take.

If you find keeping evaluation logs are a chore, develop your own system. A very simple technique is to put a copy (or the first page plus a good comment sheet) of the special (very good or very bad) reports into the file for that reason. Then, you can quickly relive the hills and valleys of the year. Above all, understand that failure to have a system makes you prey to several human failings:

(1) You will be reluctant to come to grips with poor performance. Rather than risk recriminations over items you cannot support, everyone in your group will wind up being a splendid average. Don't be surprised when the best people leave your organization and you wind up with a mediocre operation.

(2) You will try to play the good guy and upgrade everyone eventually winding up with everyone being a splendid performer. Tenured college professors can get away with grade inflation because the student who gets an A doesn't gripe about the course. The manager who inflates the group and then awards salary increments accordingly inevitably discourages the few who are truly high performers—who soon leave. The company again ends up with a (highly paid) mediocre group.

(3) You try to play the tough guy and make it next to impossible for a man to win top ratings. Again, the best men can only protect themselves by drifting out of your group (or worse, retiring on-the-job).

In all three cases, don't be surprised when your appraisal of the group as a whole leaves you with the feeling performance is at the lowest common denominator.

The manager has the opportunity to set the tone of the group; if he demands high performance and seeks results, the pressure will be on those who either cannot or will not shape up and they will drift out. Those who take pride in their work will thrive in an atmosphere of success and the feeling that they are meeting stiff standards with good results. Even if conditions beyond the control of the manager are not that good, people who are doing work they like and for a manager they know respects good work will be reluctant to move. Thus, they will become the nucleus of a top rank, respected engineering group and the manager will have good reason to support demands for better working conditions, challenging assignments, and professional treatment. If you are a manager and it is known throughout the firm that your people are interested and can solve other people's problems, that you run a very lean, but tough operation that gets the answers, you will soon find that your influence will grow where it counts.

GROUP PERFORMANCE MEASURES

There is a tendency for managers not to spend time evaluating the group as a whole. The feeling is that individual evaluations are sufficient and the fact must be faced that group evaluations carry the overtones of a manager's self-evaluation as well. The first time a manager is called by his/her superior and asked to evaluate the *group's* performance, with substantiation, it is liable to be highly embarrassing. The manager will be responding in terms of how individuals are doing, a micro view, while executives want a macro view—how the overall activity is going. Some managers never learn how to prepare for a macro examination and these executive briefings become dreaded events.

On a macro view:

(a) Group interaction—the degree to which members stimulate each other, render aid, or information is shared.
(b) The image, or prestige ranking, held by people outside the group, of the group.
(c) Executive feelings about the group and the image it projects.
(d) Interaction with other groups (complaints, successes, number of contacts, etc.).
(e) Incoming queries, requests for service, repeat business, intra-organization services you perform.
(f) Cost and schedule overruns.
(g) Project backlog.
(h) Projects completed, reports written, presentations made.
(i) Difficulty of projects assigned, innovation achieved, problems solved.
(j) Library facilities and library use.
(k) Technical depth (technicians assigned, equipment available).
(l) Freedom to attend conventions, call on outside lecturers, purchase training.
(m) Actual versus budgeted costs. Actual versus planned accomplishments.
(n) The image and feedback expressed by the clients of your group whether they are in outside agencies or internal to the firm.
(o) Significant patents, papers, creativity, breakthroughs, articles, meetings, new ideas reflecting on the group as a whole.

Again, keep a log of significant events in any of the above areas from which to amass a briefing period over-all report.

Visit conventions, work in professional organizations, and interact with other managers to refine your system through adaptation of their ideas.

DIFFICULTIES IN PERFORMANCE-VALUE DETERMINATION

The analysis of output which is the result of a mental process can be judged only on those items tangibly produced. Trying to discern what is in a person's mind and calling increased knowledge "production" is looking for trouble. If you are paying for increased knowledge, budget accordingly. But if you need a working circuit, some prototype, however poor, must be visible to assure yourself that the search for a satisfactory design is under way. The exact nature of the item to be displayed at a particular stage of product is something the manager and the subordinate must thrash out together. Avoid:

(1) Inflating estimates of progress; they lead to cost and schedule overruns.
(2) Confusion of activity with achievement.
(3) Acceptance of excuses, which stir up your negative "gut" feelings, in trying to be a nice guy.
(4) Performing not in accordance with the management's needs simply because they fail to spell out all the details. It is your duty to dig and find out the correct thing to do.

SUMMARY

A manager need not get caught unaware by changes in the firm's financial structure. Readily available financial data can often point the way to intelligent engineering initiatives.

To sustain and capitalize on these initiatives, the manager must find a method of evaluating the performance of his people. He must see that this method is acceptable to them and does not result in "working to the measure" distortion or neglect of all else. Then, these performance measures should be regularly logged—so that evaluations are intelligently performed for subordinates and the group as a whole.

The manager who continuously evaluates his "worth" to the firm, the "worth" of subordinates, and the "worth" of the group or department will find his thinking in line with that of top management. The result should be a more prosperous firm.

CHAPTER 5
Red Tape and How to Deal with It

Chapter 5
Red Tape and How to Deal with It

INTRODUCTION

Red tape is an epithet denouncing excessive proceduralism which delays and frustrates people. The term originated from real red tape used in European government organizations to tie official documents—many of which have long been challenged as unnecessary by those who prepare them.

Red tape in an organization may be real or fictional. It is real when the procedures or policies in question are dysfunctional—they impair the performance of the organization with few if any offsetting benefits. Red tape is fictitious when those who perform the procedures do not know why they are doing them; consequently, they think it is red tape, but from a broader viewpoint the work maybe both necessary and worthwhile.

UNDERSTANDING RED TAPE

ORIGINS AND USE

The epitome of red tape organization was developed under the French kings in the 18th century and was probably a reaction to what was (later) expressed as the "Iron Law of Oligarchy."[15] Briefly, it is impossible to keep everyone informed no matter how much the members desire democracy.

[15]*Robert Michels,* Political Parties, *a study of German Socialist Parties, 1915.*

Only a few can monitor the information flow and know enough about what is going on to participate in the decisions of the organization. The few who occupy the critical communication points emerge as the leaders who make decisions that affect others; they guard and control the communications points to retain power.

Under the French, the jealous guard of the king's power required the promulgation of rules and regulations to keep the king in the center of communications and to maintain his control over the realm.

FUNCTIONS OF RULES

Red tape and bureaucratic rules have become synonymous. The functions of these rules have been studied extensively and their role in modern organization is delineated by A. W. Gouldner:

(1) Rules are a functional equivalent for direct, personally given orders. These rules specify what is expected of the employee, what is proscribed, and how to carry out procedures which the organization wants followed in a set manner. Further, rules act to inform the subordinate of what bounds the hierarchy places on his discretion and the results expected within those areas of discretion.

(2) Rules perform a screening function. The supervisor, in effect, has been relieved of the necessity for indoctrinating, informing or reminding workers of particular orders. Reference can be made to the rules and the employee placed under the obligation to learn and abide by them. Since rules take over much of the onus for limiting an employee's actions or discretion, the supervisor's contact with subordinates has been screened; more attention can be given to interpersonal or project matters within the same contact time frame. Screening works both ways. It allows supervisors to impersonally enforce certain modes of behavior while allowing the employee freedom to submit "to the organization" where personalities might otherwise render that awkward.

(3) Rules permit remote control. Properly established, they enforce a maintenance of communications channels which keep "the king" informed of all organizational commitments. The rules are public; deviations are glaring. The people who deviate cannot claim ignorance of their obligations and, therefore, generally conform without the need for close supervision or continuous monitoring. Under remote control, a standard practice or procedure can be followed by engineers remote from headquarters or the deviations can be detected by the "out" group.

(4) Rules legitimize punishment. An engineer who is sanctioned for failure to conform to written company rules may feel far less the victim of unjust harassment than one who operates in an environment of ad hoc adjustments of what is or is not expected. Rules channel aggression, providing permissible avenues for its expression and legitimating the utilization of punishments.

(5) Rules provide a leeway function in that they may or may not be enforced. Formal rules are relaxed as part of an informal bargain with the employees and tightened up under periods of adversity. Status, procedures, and "channels" will be deemphasized when an organization is well on its way to meeting a tight project deadline; management has effected a bargain with its engineers—meet the challenge in return for relaxation of the rules. However, an organization under distress is frequently characterized by increasing centralization and enforcement of the rules.

(6) The status quo function of rules applies. Rules do not modify attitudes toward work, but do provide important guidelines for behavior (outwardly visible actions). Too often, rules provide a guideline for the minimum behavior an employee must exercise in order not to lose his job. It is this aspect of "red tape" which most infuriates managers seeking to motivate subordinates. Where written objectives for results do not exist or are vague, the employee who is performing subpar may compare his efforts with those required by the rules and seek to avoid sanctions.

Formal authority attaches to a position; informal power attaches to a person. An organization staffed by persons unable to combine the two will promulgate red tape to assure itself of control. This is a major origin of red tape in the modern organization.

Highly complex projects involving considerable risk to the firm cause the creation of red tape. Elaborate rules and procedures protect the organization from committing itself without at least some due care in considering costs vs. benefits.

The new manager who has just become part of the "organization" and is still feeling his/her way around may find it expedient to take shelter behind rules and regulations until the "confidence" level builds up. In such a situation, the new manager is taking advantage of the "protection" being given to the organization by red tape.

The specialization of function creates red tape. Management must secure cooperation among departments and assure itself that competing demands are properly balanced with the resources available.

These origins alone should amply demonstrate that red tape is not to be viewed solely in its dysfunctional aspects, but rather from a broader aspect. The general conception of red tape as buck passing, delay, and organizational veto needs to be examined (our first, foregoing, stage of examination was the review of its origins). Our later stages include a review of the modern organization, rational methods for dealing with red tape, and reform of the organization.

FUNCTIONS/DYSFUNCTIONS: WORD DEFINITIONS

Functions—Acts or the results of acts which confine or adjust to an organization's operations to plan. Red tape is usually credited with helping to keep an organization on plan or to function properly.

Dysfunctions—Acts or the results of acts which lessen the adjustment or adherence of operations to plan. Red tape has many effects not helpful to the organization. These are dysfunctional effects.

THE HIERARCHY OF RULES & REGULATIONS

The shape of the organization is determined by the interplay of external and internal sets of forces. These tend to generate a hierarchy of guidelines. Each guideline by itself probably has a good and sufficient reason for its existence, but collectively they serve to restrict the areas of discretionary action by the firm's personnel. These guidelines take various forms from broad policy to detailed instructions.

Listed by increasing degree the following are constraints on the manager's latitude for decision making.

Policy—The principles on which courses of action are based. Serve to give broad guidance to those setting the firm's goals and how to reach them.

Procedure—Specification of a particular course of action or modes of action by which to accomplish the tasks of the firm.

Regulation—Boundary not to be exceeded in performance of duties of the firm. Also interpreted as a "rule" prescribed by a superior.

Rule—Procedural question answered by those in authority to assure that future situations yield the same responses.

Method—Clearly defined way to reach an end. Also interpreted as procedure characteristic of a particular discipline such as engineering, mathematics, etc.

Instruction—An order or mandate for specific actions in a specific sequence at a specific time.

EXTERNAL FORCES Consumers, clients, government, and other outside forces tend to generate a hierarchy of policies, rules, etc., which is adhered to and enforced more rigorously than others because of the following reasons.

(1) The organization must protect the full spectrum of its interests in the long as well as short run. It cannot allow transitory considerations or expediency to conflict with the broader ramifications of what it is trying to accomplish; it protects itself by the conscious creation of red tape. A manager who deals with outsiders, whether they are customers, government personnel, suppliers, or even competitors must be aware of pitfalls. Thus, the firm establishes detailed procedures for response.

Examples

Dealing with customers—An engineer making an uninformed recommendation of a product can expose the firm to a serious product liability suit should the product not "fit" and cause damage.

Dealing with government—Unwise statements or irresponsible actions open the firm to regulatory action, investigations, or simply adverse publicity.

Dealing with suppliers—If your firm is a major marketing force your innocent inquiries may be interpreted as coercive, attempts to exert undue influence, etc.

Dealing with competitors—Innocent cooperation may be interpreted as industry collusion or monopolistic practice.

(2) The organization wants a consistent response from any component part to any outside inquiry. At one extreme, bureaucracy will direct everything from the outside to a central response center. Here, all incoming matters will be screened, responsibility for preparing responses will be assigned, and the prepared responses carefully filtered before being sent out.

Example

An "innocent" inquiry to the engineering department may mask an attorney's probing for weaknesses in a product liability suit. Thus, policies and procedures forcing a central legal review may not only provide protection, but prevent the outsider from probing different people over a period of time to build a case.

(3) The organization is responsible to a multitude of interested outside agencies, societies, owners, and other firms. It cannot allow its infor-

mation channels to become unbalanced so that releases to one are inconsistent with releases to the others. The inquiries or demands made to a firm have become extremely complex; they frequently require the combined efforts of a multitude of departments to appropriately respond. Red tape will be intentionally established to be sure that the information is screened and is consistent from channel to channel, and there are no "leaks" from the departments involved.

DEFINITION A bureaucracy is an organization in which work is divided on the basis of specialized expert knowledge and ability and which has superimposed on this division a highly elaborated hierarchy of authority. The advance of specialization and the elaboration of the hierarchy have yielded larger and more bureaucratic organizations.

INTERNAL FORCES The modern organization has become a bureaucracy. Red tape is used to assure control over the means when the ends are difficult to specify or the area of discretion of the employee must be limited. This in turn generates a series of internal forces which shape the organization and the activities of those who work in it.

(1) The bureaucracy stresses equal treatment and consistency of operations. The exception becomes a threat regardless of the motives behind the exception.

(2) The bureaucrat, at the extreme, has an inner compulsion to fit all things to established patterns; facts may be reshaped and situations forced to fit even when they do not. The apparatus of the bureaucracy uses rituals (procedures), symbols (forms), and a legislature (committee) to legitimize this forced fitting and ratify collective decisions.

(3) The organization will use a combination of written and unwritten rules to select and promote executives so that a general condition of compatibility of personnel is maintained. Those who are chosen are the ones who "fit." Fitness includes personal distinctions, image, personal characteristics, and appropriate background.

(4) The organization stresses rationalism—how does this support the company goal? Goals are in turn factored into subgoals which are in turn factored, etc.

(5) The organization fosters an illusion of assured careers for its personnel. This is to encourage them to invest time and effort in the specialized nature of the work the firm is doing and to build competence in that area. An "assured" career allays concern for the transferability of that specialty and reduces tensions. Specialization builds a need to protect that specialty—the organization will resist change if a specialty and the "assured career" concept are threatened without overwhelming (compensating) benefits.

(6) Finally, the manager's freedom to operate is being eroded by the shift of much of the intellectual content of planning, organizing, etc., to staff specialists. This leaves the hierarchical rights and duties as the principal components of the manager's job and these are, in turn, increasingly protected by rules and regulations—red tape.

ORGANIZATIONAL PATHOLOGIES

The most frequently voiced criticism of red tape is that it generates resistance to change and, consequently, stifles creativity and innovation. The criticism may not be entirely fair. A control system preventing decision by expediency or precipitous changes in the business of the firm will be labeled red tape by disappointed proponents of the changes—but reviewed as protection against foolishness by the hierarchy. Resistance to change is not unique to the bureaucracy and is usually the first organizational pathology to cause the creation of red tape.

Most organizations develop an attitude of caution before taking any action which might disrupt the management operation, the work flow or the technical specialization of that work. Part of this attitude grows out of ignorance of the true nature of the organization's response to the stimulus of a new idea or proposal. The organization, they fear, may be underdamped; a change, or transient input, may create oscillations they cannot anticipate and which go out of control. The managers fear for their personal survival in that organization. On the other hand it is far more likely that the organization is overdamped with only a few people aware of the dangers this presents to the long term survival of the firm. A change, even when desperately needed to meet new circumstances, is buried under the system and its red tape, and from proposal to execution creates scarcely a ripple in the day-to-day operations. Life goes on as before; any external change re the markets, the economics of industry, the law, or other influences spells a major crisis for such an organization. The ideal, of course, is a critically damped system in which ill-considered or unimportant transients are "absorbed," while other signals do cause an appropriate response.

At an early stage of an organization's existence, an engineer can walk into the vice president's office, sketch an idea on a pad, and walk out with a decision. A mature organization requires project forms, profit analyses, and capital funding projections. The difference is not a hardening of the "change arteries" or intent. The organization does want to make a profit, it does want to stay competitive and up to date. At maturity, however, it has more and different projects to consider. Owners, investors, and the people of the corporation are past the do-or-die, win-or-lose risk acceptance stage; the stakes are now higher.

A second pathology is more serious. It involves a means–ends reversal and a confusion between needs and wants. For example, the paper statements of goals are not ends; they are means by which efforts are appropriately focused. The ends are the products moving out of the door or, viewed on a higher plane; they are the dividends flowing to the stockholders or the increase in stockholders' equity. Similarly, the means will be mistaken for the ends—wherein the establishment or adherence to procedure becomes primary and the ends are lost. A good manager is quick to detect this reversal and works to recreate a proper goal orientation in the organization.

The last item in our brief survey of organization pathologies caused by red tape involves the shifting of the costs of the system by those with more authority down to those with less. The obligation to accept another's decisions carries with it a number of negative aspects or costs. The decision may not be in accord with one's technical knowledge or with one's perception of the opportunities available, may conflict with one's self-interests, or may require a change in behavior. The organization allows the persons higher up in the hierarchy to satisfy more of their personal needs than it does those on the lower levels in the scope and impact of the decisions made. The new manager frequently is startled at the impact of these costs. Attempting to minimize them by forcing a viewpoint upwards (reversing the cost shift) causes the person involved to be labeled a "maverick." A manager who fights for what he/she believes is "right," may cause such a degree of shift reversal that management brands that manager a failure. "We have taken a good engineer and made a poor manager."

Shift reversals can be downplayed by practicing "good interpersonal relations." Be aware of what is happening; speak to people, act towards people as you would have them act toward you when you are in an uncomfortable situation. Also, be absolutely certain that what you espouse is "right." Right means correct. It means that you have done your homework "to the limit," all your calculations have been double checked by alternative methods, all your assumptions are clearly stated and withstand impartial examination, all the information sources have been carefully checked, all the people who will be impacted know what you are doing, and the matter withstands every "acid test" management has been known to apply. If you are attacking the test itself, your need to be correct is multiplied. It is surprising how many times people overlook obvious points when they become emotionally involved in furthering an idea. The first person who takes an active interest in the idea (or wrecking it because he does not like the cost shift) will search out this overlooked fact and deal with the proposal accordingly.

PERSONAL PATHOLOGIES

The new manager in a bureaucratic organization lives with anxiety and tensions. Dependent on subordinates to achieve technical goals, the manager is, nevertheless, the one who will be called on the carpet in the event of failures. To protect themselves managers make the employees comply with the control devices and will amplify, adjust or redefine goals, policies, and procedures as necessary to avoid failure. Some subordinates, in turn, view the ''game'' as avoiding black marks. Decisions by precedence are deemed safer than initiatives, and sitting tight to avoid rocking the smoothly functioning organization is the modus operandi. The manager seeking to ''motivate'' the organization applies more control mechanisms, tightens the objectives, and may even introduce an element of stress. Stress and anxiety continue to build within the new manager and he creates tighter and tighter controls, more red tape, in trying to lessen them.

Red tape becomes, even more, a type of defense mechanism; the rituals of defined procedures become rituals of stress and tension release. Projects, initiatives, and any changes must, in effect, undergo the ritual ''cleansing.'' If they pass, then the manager can feel safe in implementing or otherwise acting on them. Taken to extreme, these rituals can paralyze an organization. The lubricant which prevents this is the informal organization—the personal contacts and interactions which bend the rules or their interpretation to allow a flexibility in function.

OF SPECIAL NOTE

The most important personal pathology triggered by the red tape organization, which affects the new manager, is an unreasoning rebellion against the red tape. This rebellion may damage the new manager's career more than the overt circumstances may seem to justify. Consider the following ideas espoused by ''experts'' in the study of organizational operations—how they view the rebellion personal pathology:

(1) Whether or not red tape is necessary in an organization is relative to personal values and beliefs—not some immutable standard. The person who proclaims red tape as unnecessary may have only a limited view of what is necessary to others.

(2) People who are personally insecure and tense, who have a dislike of waiting for results, have an underdeveloped power of investing energy in future objects. The frustration of immediate purpose is denounced as ''red tape.''

(3) This person will echo the complaint of fellow frustrates: "The organization wants conformers." He begins to feel that "pull" is the only thing which gets him ahead.

(4) This person does not understand, or bother to learn, why red tape exists. Consequently he violates all the rules (not only the unnecessary or excessive constraints), refuses to keep required records or make required reports—or does so in a very poor fashion—thereby confusing and increasing the workload of others.

(5) This person will express a desire for simpler, less restricted relationships, action through amicism (the pursuit of interests by means of personal contacts with strategically placed peers), and a return to the good old days in which the corner grocery store represents the appropriate model of no red tape. The demise of the corner grocery store to that epitome of red tape—the chain store—carries too little weight with this individual.

(6) Top management views such feelings as convenient rationalizations for lack of personal success. They are viewed as alienation and call the person's loyalty into question.

(7) This person is also viewed as immature: A person's ability to wait grows stronger as he matures. Frustration and rebellion are seen as an underdeveloped ability to act as a mature adult. It is a failure to understand that the large complex organization cannot make an exception in his case as it moves forward with a certain deliberateness.

(8) When a person criticizes the organization requirement as red tape, it is probably just as appropriate to investigate the person so reacting as it is the organization.

In summary: The experts tell us that when a person develops the pathology of rebellion against red tape, the problem, quite often, is more with the person than with the organization. They tell the organization, or rather support it in the view, that the rebellious individual is one to be given careful "examination" and not always a sympathetic hearing. (Before rebelling, consider the later section in this chapter on how to reform red tape.)

THE RATIONAL APPROACH TO RED TAPE

THE PERSONAL APPROACH

(1) Understand the origins of red tape, the forces shaping the organization's view of rebellion against red tape. This is your general information base from which your ideas for action and criteria for evaluation (particularly the consequences) of those actions are generated.

(2) Understand that the most pernicious trait a manager can display when faced with obstacles is indecisiveness. Very often this is explained as a need for mature consideration of the situation, but actually comes from the mind slipping away, from a tension generator, into inappropriate or trivial matters. The obstacle hurdler makes decisions when he/she can—and the sooner the better. The obstacle breaker has a high tolerance for frustration, does not blow his stack, and works around delays, committees, and clearances with Spartan endurance.

(3) Study and flow chart the communication channels, the work flow channels, the command and control channels, and the red tape which governs these. This is your "specific information base" from which specific tactics for operating in your organization are shaped. Establish, in effect, a personal operating manual.

(4) Analyze the key positions in the organization vis-a-vis your own advancement in the organization. Analyze the occupants of those positions. Then become a key component of their informal networks by:

a. Protecting them from their own and others' mistakes by learning of their problems and goals.
b. Providing assistance where it can be volunteered.
c. Providing a source of expertise—a known information source in the person of yourself.
d. Projecting an image of cooperation with all phases of the firm's operation at all levels.
e. Helping them absorb the costs of red tape and relieving stress.

(5) Establish a system of check lists for maintaining contacts (touching bases) with:

a. Persons whose opinions should be solicited even if only as a courtesy.
b. Persons whose clearance and approval should be solicited even if only as a courtesy.
c. Persons who should be kept informed even if only as a courtesy.

(6) Make a checklist of the appropriate channels, routes, and procedures to follow for each class or type of action/communication you initiate.

(7) Build a personal network of informal contacts to lubricate the formal organization.

COMMENT

Building a personal network is more than selling services. Selling stresses the outsider trying to get in. You should form social and technical contacts in which you are the one called. Then you become the insider to the persons who have pre-sold themselves on the benefits of knowing you. Just as in dealing with the occupants of the key positions in the organization, this is a process consisting of careful and deliberate steps.

THE TACTICAL APPROACH

George Strauss[16] has identified five tactical methods you can use in dealing with other departments.

(1) Rule Oriented Tactics

a. Appeal to a common authority when it is necessary to alter the nature of a request.
b. Refer to some policy or procedure which gives you a favorable upper hand.
c. Require the department seeking the service to absorb on its budget all the costs of your service.

(2) Rule Evading Tactics

a. Go through the motions of complying with the request, but show no results.
b. Exceed formal authority and ignore the request altogether.
c. Acknowledge the request, but assign it a sliding priority which is always lower than all other backlogged work.

(3) Personal–Political Tactics

a. Rely on friendships to achieve approvals, decision, action.
b. Rely on favors, past and future, to achieve the same result.
c. Work through political allies in other departments to bring pressure on a targeted individual.

[16]*Quoted in Leonard Sayles,* Managerial Behavior: Administration in Complex Organization, *New York: McGraw-Hill, 1964.*

(4) Educational Tactics

a. Use direct persuasion to convince people of the rightness of your cause. Persuasion backed up by completed staff work can be very effective when combined with political tactics.

b. Use indirect persuasion over a long-term period to help others see things from your department's point of view.

(5) Organizational-Interactional Tactics

a. Seek to change the interaction pattern—have your department placed under a different classification or position in the work flow/communication channel.

b. Seek to take over other departments.

These tactics are also called bureaucratic gamesmanship.

REFORMING THE ORGANIZATION

OVERVIEW

Reforming an organization is a job that is beyond the control of lower level managers. The manager who wishes to go beyond coping with the red tape to eliminating or reducing it should aim for small increments of change, over a long period of time, based on detailed, comprehensive thought on the matter (i.e., cost/benefits, consequences, risks, etc.).

BASIC PRINCIPLES OF REFORM

(1) Build your general and specific knowledge bases about red tape.

(2) Gain solid, first hand experience in complying with the red tape before seeking change.

(3) Be careful of whose ox will be gored by any change from the existing procedures. If a superior has a vested interest a change will threaten, act accordingly.

(4) Touch bases with all persons interested or impacted by a potential change while the change is still in its architectural state—not a frozen, finished, typed document. (A "neat" trick: show people sloppy rough drafts when asking for comments—not a finished piece of work.)

(5) Work through your superior—not around him/her. Share your ideas and authorship with your superior. Such joint equity will commit the superior to success of the change and reassure him of your loyalty—a key requirement for advancement. If your sale of equity fails and the superior does not give permission to take the idea further, consider strongly that an organization's own first loyalty is to protect the hierarchy—not you, no matter what you are selling.

(6) Your idea must be constructive, fully evaluated as to impact, costs, benefits, external pressures, and with the approval of all who are impacted. Very often, all that is needed is the absence of objection, like the Quaker's decision process. Your presentation must embody the principles of completed staff work. In particular, check with the firm's legal department to see if there may have been a reason behind the particular piece of red-tape.

CUTTING RED TAPE

Proceeding without authorization to "get the job done" is highly touted in the literature and on Broadway. In those instances the hero succeeds and the firm climbs to new heights of sales, etc. The failures are quietly buried by the wayside and don't quite make the headlines.

However, the question does arise: what is one to do when the red tape process interferes with an important and time constrained assignment? We offer the following suggestions:

(1) If you are cutting through tape, maintain the attitude, publicly, that you are doing it for this job only because of time, importance, etc. Don't let yourself be branded as a rebel against the system; the mere fact that you are now cutting the tape will then be proof positive you are a rebel. The system is set up to expedite routine matters. For exceptional, or nonroutine matters, special treatment is "required."

(2) Work through your superior and get tacit approval for what you are doing. If what you have done turns out to be less than brilliant, the post-mortem examinations will at least show that you respected the hierarchy.

(3) Touch base by telephone and inform everyone possible about what you are doing "this one time only" because of such tight time and importance constraints. Whether you get or need their approval or not, the fact that they didn't give you a flat cease and desist order morally commits them to some responsibility for the results—and to defend you accordingly.

(4) Make absolutely sure you have all the facts: Is the item really important to all concerned? Or is it your interpretation of importance, on your scale of values, which might be different from the values of others?

(5) Your group's output is someone else's input. If you bend the rules, don't increase someone else's workload without their approval, or have a good explanation ready before they are hit with the work, not after.

(6) Evaluate carefully the "upside" rewards and the "downside" costs in dollar terms, in terms of time, of your reputation as a manager in the system, and of just what you will accomplish—also in terms of future difficulties your project will incur (known as winning the battle, but losing the war). If all this makes you stop and do a little paper analysis and plan before you rush into battle, your chances of success will have been increased markedly and downside risks reduced. Make absolutely certain your goal or objective is clear cut; if a superior were to ask you what it is, hesitating, thinking on your feet, or voicing innocuous vacuities will be interpreted as fuzzy thinking and project a terrible image of you.

(7) Share the benefits. When things turn out "right," praise the people who helped you even if all they did was to refrain from interfering when they could have. In academic circles, it is not unknown for two authors working separately on their books to jointly list each other; the feeling is that it is better to claim authorship of *two* books, albeit on a shared basis, than sole accomplishment of *one*.

(8) Keep people informed about the goals and benefits of the work you are doing. If you see this chore as personal public relations (PR) the doing might not be as onerous.

(9) Develop a "standing plan" for yourself as to just how you would cut red tape should the occasion arise. A standing plan is one made in advance of actual need for it and kept at ready for the correct circumstances to arise. The plan need be no more than a checklist of people to call, things to say, reports to frame, etc. Such a plan prevents you from "shooting from the hip."

(10) Be conscious of the fact that the ability to cut corners comes from power you have "banked" in the minds of people re your:

a. Expertise
b. Effort and interest in your work
c. Record of achievement
d. Personality
f. Knowledge of the rules
g. Ability and willingness to return favors.

When you cut corners you draw from that bank.

(11) Employ a "political" approach where it may yield success. In a political situation, you are attempting to "get some other people to act or stop acting in a certain way in order to achieve a goal important to you." Simply stated, you are trying to get some people to stop what they are doing and to get others to do things they might not want to do. To be successful, you will want to:

a. Probe your surroundings to figure out who are the most important "actors."
b. Interact with them to find out their true inclinations and their influence on the items of importance to you.
c. Calculate how to get them to behave the way you want them to behave.
d. Execute your plans.

You may want to set up detailed charts showing exactly who is important or not in the organization and their degree of influence over matters vital to your success; you will then find opportunities to get close to them to get your readings of their inclinations. (Eugene Jennings calls this "visiposure"—a combination of visibility and exposure to the important people to project an image and to gain feedback.) Your strategy for influencing their behavior may then include alliances, co-opting, etc. [See Coplin/O'Leary, *Everyman's Prince,* Duxbury Press, North Scituate, Massachusetts, 1976, Chapters 1 and 2.]

STRATEGY OF THE SINGLE REPORT

The intelligent elimination of paperwork is one way to cut red tape if careful attention is paid to all the other considerations bounding the red tape. The principle of the single report may be useful.

(1) Make an inventory of all the reports your group prepares.

(2) Take an intense critical look at what the reports do and don't do.

(3) Determine the nature of the decisions being made by the user of the reports and the information extracted by all who are exposed to them.

(4) Examine the reports for duplicate material or material showing only slight variations from one to the other.

(5) Cut and paste sample blocks of information so that all different pieces are combined into one report format.

(6) Develop a multitude of sources and channels for information flow to

you from subordinates, peers, clients, superiors, vendors, etc. Cultivate, especially, those who can supply information such as planning, accounting, personnel, and those who will be impacted by your decisions.

(7) Combine material so that it shows appropriate criteria for the decisionmaker:
   a. *Single criterion*—only one quantity is measured and observed for decisions.
   b. *Multiple criteria*—several quantities are measured simultaneously such as output, cost, safety, quality, etc., for decisionmaking.
   c. *Composite criteria*—separately measured quantities are weighted and combined into indices or other indicators.

YOUR OWN RED TAPE

In all of the foregoing discussions it must never be forgotten that the ultimate payoff to your firm is derived from the competence of your group. In the words of an engineering vice president, "no firm will exist for very long where the engineers cannot turn out good work." He emphasizes that the manager's first duty to the firm is to see to it that what comes out of his department is correct and of highly competent design. If the group cannot be implicitly trusted to put out a uniformly high standard of work, the manager is not doing his job. Some suggestions:

a. Maintain a system for cross checking from one engineer to the next the quality of the decisions made, the correctness of specifications drawn up, or designs to be approved.

b. If necessary, split the group and run parallel calculation exercises to see if both arrive at the same answer.

c. Maintain your own technical competence at a level where you can cross check your engineers' outputs. Make it a habit to take home a sampling of the work output documents for a detailed "going over." Where a matter is beyond your technical competency, find a peer supervisor and get his/her help (and be ready to return the favor). Note: Such peer supervisor crosschecking of the work is a superb way of building alliances in a firm and moving up together.

d. Where a lot of money is "riding" on your output, ask the firm to let you bring in a free-lance consultant to do some parallel work, to draw some independent conclusions, and to then independently review the output of your group.

TABLE 5.1

Format of a Report Representing Completed Staff Work

I. Synopsis:
   (a) One sentence on why report was written.
   (b) The major conclusion in a few sentences or less.
   (c) Action recommended.

II. Problem or decision formulation:
   (a) Issue of concern is clarified and all assumptions stated.
   (b) The objective behind the work—what it was supposed to accomplish.
   (c) The objective of this report.
   (d) Definitions and delimitations of the problem or decision situation.

III. Search sources and methods used:
   (a) Credibility of sources.
   (b) Data gathered.
   (c) Limits of search.
   (d) Models used.
   (e) Organization of the data.

IV. Analysis methods used and justification for their use.

V. Evaluation of consequences and follow-on actions:
   (a) Tangibles and intangibles which did not "fit" the analysis but which bear on situation.
   (b) Probabilities that affect risks in project.
   (c) Risks—upside and downside payoffs.

VI. Interpretation and researcher's opinions. Items which, again, did not "fit" into the standard analyses performed, but which are worth mentioning. Impressions, ideas, suggetions beyond the immediate scope or report fit here.

VII. Conclusions in full. These follow from items II, III, IV, and V.

VIII. Verification:
   (a) How the work might be checked by the next level up in the hierarchy.
   (b) How the work should be evaluated.

IX. Actions to be taken:
   (a) All goals and objectives for these actions.
   (b) All plans and subplans.

X. Appendix:
   (a) The raw data.
   (b) Computer output (synopsis if long).
   (c) Synopsis of programming used (if any).
   (d) All other detail material.
   (e) References to other internal documents and reports.
   (f) References to other books, externally supplied/available material.

e. Train and retrain your people. Check the courses they are taking, be concerned with the technical matter they are learning, and review their grades. Maintain an atmosphere of technical competence and quality of output.

f. Give your subordinates sample formats and examples of the type of work you want them to emulate—where possible. Consider, for example, the format of a report to management shown in Table 5.1.

If your subordinates rebel at the checking red-tape and report formats you create, consider carefully the downside risks of a retreat.

You may have little to no control over the mass of coordinators, expediters, planners, and paperwork which afflict modern organizations. However, within your group you can maintain the advantages of small size: make sure people talk to each other, stay current in their specialties, reasonably competent in the specialties of the other group members, are recognized for their technical ability, and shield them as much as possible from the red tape you handle.

CONCLUSIONS

(1) Red-tape may not be a pathology of the organization but rather a response to the forces shaping it and a form of self-protection.

(2) Working through it indicates a high degree of tenacity and maturity. Cutting through it is not always beneficial to the firm or the image of the cutter.

(3) Successful reformers of red tape are those who build a successful performance record within the system, who have clearly delineated goals for the changes they seek, and perform "complete staff work."

CHAPTER 6
The Manager's Basic Bookshelf

Chapter 6
A Manager's Basic Bookshelf

The rationale behind the choice of the mentioned books is that they provide new or basic information in their subject areas. At the completion of the books listed in an area, you should have enough knowledge to pick and choose particular books for your own follow-on reading.

HUMAN RELATIONS & PSYCHOLOGY AREA

Argyris, Chris, *Personality and Organization:* Harper & Row, New York, 1957. The classic in examining the growth and maturity of the individual versus the forces at work to keep the person from true growth on the job.

Jennings, Eugene E., *Routes to the Executive Suite:* McGraw-Hill, New York, 1971. An applied psychology and empirical study of what it takes to move up in the organization.

The following books are good basic reading in the field:

Leavitt, Harold J., *Managerial Psychology:* University of Chicago Press, Chicago, IL (paperback), 1972.

Leavitt, Harold J., and Pondy, Louis R., Eds., *Readings in Managerial Psychology:* University of Chicago Press, Chicago, IL (paperback), 1974.

Levinson, Harry, *Exceptional Executive:* Mentor Executive Library (paperback), 1971.

Sayles, Leonard, *Managerial Behavior: Administration in Complex Organization:* McGraw-Hill, New York, 1964. One of the finest behavioral analysis books; depends heavily on empirical research giving it meaning to practicing manager. Not an easy book to read—demands attention and rewards accordingly.

THE MANAGEMENT PROCESS (PLANNING, ORGANIZING, ETC)

Items (1), (2), (3) are basic textbooks in the field for comprehensive reading.

(1) Dale, Ernest, *Management: Theory and Practice:* McGraw-Hill, New York, 1965.

(2) Karger, Delman, and Murdick, Robert G., *Managing Engineering and Research:* Industrial Press, Inc., New York, 1969.

(3) Terry, George R., *Principles of Management:* Richard D. Irwin, Inc., Homewood, IL, 1972.

Fayol, Henri, *General and Industrial Management.* Pitman Publishing Corp., New York, 1949. The basic classic in this field, 96 pages long, from which all other writers seem to have copied. Out of print. Obtain from used booksellers.

Kazmier, I., *Principles of Management: A Programmed-Instructional Approach:* McGraw-Hill, New York (paperback), 1974. A self-learner which combines the process of management with behavioral aspects. A reader taking this programmed course will have completed an extremely solid background in management.

Parkinson, C. Northcote, *Parkinson's Law:* Houghton Mifflin Co., Boston, MA, 1957. The classic spoof of the organization with, however, basic teachings.

ECONOMICS

Random House (New York,) puts out a series of paperback "primers" written on a sophisticated level in a down-to-earth manner. A person can gain a high level of understanding even when starting from "ground zero" in knowledge in particular areas with each of these short books—all paper and about $4.00 or so in cost each. Three of the basic ones management should understand are:

(1) Bernstein, Peter L., *A Primer on Money, Banking and Gold* (paperback), 1969.

(2) Heilbroner, Robert L., and Bernstein, Peter L., *A Primer on Government Spending* (paperback), 1970.

(3) Pen, Jan, *A Primer on International Trade* (paperback), 1967.

Heilbroner, Robert L., *The Wordly Philosophers:* Simon and Schuster, New York, 1972. Hilarious description of the lives of the great economists; author teaches the most important economic theories each man developed over the last several hundred years.

Hitch, Charles J., and McKean, Roland N., *The Economics of Defense in the Nuclear Age:* Harvard University Press, Cambridge, MA (paperback), 1960. Must reading for anyone in government (or industry as well) making a decision or solving a problem involving costs, basic economic considerations, benefits.

Murad, Anatol, *What Keynes Means:* College and University Press, New Haven, CT (paperback), 1962. The finest book to unravel *Keynes' General Theory of Employment, Interest, and Money* and the course of the U.S. economy.

BASIC LAW

Three, easy to read, Dell Books (Dell Publishers, Pinebrook, NJ) give an excellent insight into the law for people with no background in the field.

Deming, Richard, *Man and Society: Criminal Law at Work,* 1970.

Deming, Richard, *Man Against Man: Civil Law at Work,* 1972.

Deming, Richard, *Man and the World: International Law at Work,* 1974.

MARKETING

Dibbs and Pereira, *Promoting Sales,* ILO Office: A programmed learner—handles the "selling" of a product. Concepts applicable to a manager interested in "selling" a project to higher management.

International Labour Office, *Creating a Market:* An ILO programmed book. A self-learner, goes from "ground zero" up to sophisticated concepts as you work in the book and gain a basic knowledge of marketing. Order from ILO office shown under accounting.

ACCOUNTING

International Labour Office, *How to Read a Balance Sheet:* An ILO programmed book. The best method of learning basic accounting at the price of a paperback. Can be ordered from:

ILO Branch Office
1750 New York Ave. N.W.
Washington D.C. 20006

PROBABILITY & STATISTICS

A "painless" way to learn the subject is through a programmed learner. Two of the best are shown below.

Dixon, John R. *A Programmed Introduction to Probability:* Wiley, New York (paperback).

Mosteller, F. R., *et al., Probability and Statistics:* Addison-Wesley, Reading, MA (paperback), 1961.

OPERATIONS RESEARCH: LINEAR PROGRAMMING

Buffa and Dyer, *Management Science/Operations Research:* Wiley, New York. A basic textbook in the field.

Bursk, Edward C., and Chapman, John F., Eds., *New Decision Making Tools for Managers:* Mentor Executive Library Paper, (paperback), 1971.

Metzger, Robert W., *Mathematical Programming:* Wiley, New York (paperback). Leads you by the hand through operations research linear programming methodology and "into" the field in an interesting, applied manner.

PROBLEM SOLVING & DECISIONMAKING

Lyden, Fremont J., and Miller, Ernest G., *Planning-Programming-Budgeting: A Systems Approach to Management:* Markham Publishing, Chicago, IL, 1972. Don't be put off by the accent on PPBS in the title. If a manager's job in solving problems and making decisions involves the systematic reporting to top management, the principle of completed staff work encompasses "a systems approach" to these decisions, etc. This is the book for learning how to be a systems analyst (in the sense of being complete and in the sense of being thorough in making management decisions).

Martino, Joseph P., *Technological Forecasting for Decision Making:* Elsevier Press, American Elsevier Publishing, New York. One of the newest and finest books on forecasting using Delphi, analogies, etc. Shows how to make forecast and how to apply it to every day work including a major section on planning R&D. Paperback: $19.50.

Odiorne, George S., *Management Decisions by Objectives:* Prentice–Hall Inc., Englewood Cliffs, NJ, 1968. A basic, excellent book on cost-effectiveness analyses and all their ramifications.

Richards, Tudor, *Problem Solving Through Creative Analysis:* Halsted Press (John Wiley subsid.), New York, 1974. The basic book which examines the creative thinking process and methods of "cracking problems," working with decisions, etc.

MANAGEMENT BY OBJECTIVES

Mager, Robert F., *Goal Analysis:* Fearon Publishers, Lear Siegler, Belmont, CA (paperback), 1972. Tells how to write "non-fuzzy" goals.

Morrisey, George, *Management by Objectives and Results:* Addison–Wesley, Reading, MA (paperback), 1970. The book on the subject is applied enough so a manager can use system.

CAREER PLANNING

Glaser, Barney G. *Organizational Careers: A Sourcebook for Theory:* Irvington Publishers, New York, 1968. An extremely thorough collection of articles combining theory and empirical research on people's careers in the white collar fields. Everything from the movement and progression of the engineer to studies of the manager.

Author's Biography

Irwin Gray is currently president of his own firm, Envort-Gray Corporation, which is a manufacturer and distributor of energy conservation controls and systems. He is also a member of the business management faculty of New York Institute of Technology in Old Westbury, Long Island.

He holds the B.E.E., M.S., M.B.A., and Ph.D. degrees. He is a licensed P.E. (New York).

Dr. Gray spent 10 years in industry as engineer, group leader, and supervisor with such firms as Sperry, Kollsman, and Pan American World Airways. He has been a professor of business management at Pratt Institute's School of Engineering, the Polytechnic of New York, in addition to his current position at NYIT. He has done extensive work in these programs in setting up development courses for newly promoted engineers.

Dr. Gray has also done consulting for industrial firms, governmental agencies, and non-profit operations in the fields of decision making, budgets, operations, and managerial development. His specialty in training is the development of newly promoted personnel into positions of managerial responsibility.

Dr. Gray has published articles dealing with worker skills, development of managerial personnel, management by objectives, and the engineering/management aspects of product liability. His previous book was for AMACOM, on products liability.

This book developed out of Dr. Gray's work for clients and the best judgments of the members of the education committee of the Engineering Management Society. All concerned felt that the items covered in this transitions book reflect the most pressing matters facing a new supervisor making the jump from professional specialist to supervision.